CARPENTRY MADE EASY;

OR,

The Science and Art of Framing,

ON A NEW AND IMPROVED SYSTEM

BY WILLIAM BELL

ARCHITECT AND PRACTICAL BUILDER

PHILADELPHIA:
HOWARD CHALLEN
1858

The Toolemera Press

www.toolemerabooks.com

Carpentry Made Easy; Or, The Science And Art Of Framing, On A New And Improved System
By William Bell
Originally Published: Philadelphia: Howard Challen; 1858

No part of this book may be reproduced, stored in an electronic retrieval system, or transmitted in any form or by an means, electronic, mechanical, photocopy, photographic or otherwise without the written permission of the publisher.

Excerpts of one page or less for the purposes of review and comment are permissible.

Copyright © 2013 The Toolemera Press
All rights reserved.

International Standard Book Number
ISBN : 978-0-9897477-8-3
(Trade Paper)

Published by
The Toolemera Press
Massachusetts U.S.A.

Manufactured in the United States of America

www.toolemerabooks.com

Introduction
by Gary Roberts: Publisher; The Toolemera Press

William E. Bell, of Ottawa, Illinois, published *Carpentry Made Easy; Or, The Science And Art Of Framing, Or A New And Improved System,* in Philadelphia in 1858. *Carpentry Made Easy* remained in print from 1858 through 1904. Revised and enlarged by 18 pages in 1875, this reprint is of the first and most influential edition of 1858.

Carpentry Made Easy was the first 19th century architectural book to popularize the transition from heavy timber frame to light balloon frame construction. Included are the various major framing methods of the period for everything from small house construction through to barns, factories, church steeples and bridges. Although not the first book in the United States to introduce balloon frame construction, *Carpentry Made Easy* was the first technical book to thoroughly describe the method in such a way as to allow the skilled practical carpenter to readily apply this affordable building method to everyday use.

Bell described himself on the title page as an "Architect And Practical Builder", a term which at that time referred to someone who specialized in both building design and on-site construction. As noted in his biography, Bell was a trained carpenter and joiner who was experienced in building everything from small houses to churches and bridges. It was Bell's careful selection of the 126 illustrations contained in 38 full page engraved plates, each accompanied by technically precise explanations that any skilled carpenter could follow and learn from, that proved to be the deciding factor in the popularity of *Carpentry Made Easy*.

NOTE: The Toolemera Press edition of *Carpentry Made Easy* contains two additional cutout paper designs, original to a period owner of the book, between Plate 19 and Page 73.

Biography

From: History Of La Salle County, Illinois; Vol. I; Inter-State Publishing Co., pp. 508-509; 1886

William E. Bell is a native of Berkley County, Va., born June 5, 1815, a son of Vincent and Rachel (Chenoweth) Bell. In 1832 Vincent Bell moved to Seneca County, Ohio, and bought land on the Seneca Reserve, where he lived until his death in 1856. His wife died in 1858. He raised a company for the War of 1812, and was commissioned its Captain, but arrived in Baltimore after its evacuation by the British, and therefore too late to participate in active service. His grandfather, Jeremiah Bell, was a member of the company known as Washington's life-guards in the war of the Revolution.

William E. Bell was seventeen years old when his parents moved to Ohio. He was soon apprenticed to learn the carpenter and joiner's trade, working in the summer and teaching school in the winter. He worked three summers when his employer said he was as good a mechanic as ever went out of the shop. He then went to Michigan City, Ind., and went into a shop where there were thirty workmen, and was surprised to learn how little he knew about his trade.

In December, 1836, he came to Illinois and located at Tiskilwa, Bureau County, where he remained ten years, and in 1846 moved to Ottawa, where he has since lived. He was employed two years in La Salle County by Josiah Pope, his first work being on a Congregational church. For thirty years he worked for the Rock Island and Northwestern Railroad companies building many of their bridges and shops.

Mr. Bell is thoroughly conversant with every detail of his trade, and is one of the best mechanics in the State. In 1858 he published "The Art and Science of Carpentry Made Easy," which was revised and enlarged in 1875.

Mr. Bell was married in Tiskilwa to Almira Hadley, a native of Muskingum County, Ohio, born May 27, 1823, daughter of Henry Hadley. They have had six children; but three are living - William S., Frank E., and Ermina, wife of J. A. Hossick. Mr. and Mrs. Bell are members of the First Baptist Church of Ottawa.

Toolemera Press Reprints
www.toolemerabooks.com

The Toolemera Press reprints classic books, photographs & ephemera of early crafts, trades and industries, all carefully selected from our personal library.

- The New-England Farrier: Paul Jewett 1826
- The Military Exercise Of The Independent Company of Cadets: Boston 1818
- The Little Confectioner: Smith Hicks 1876
- Text Book Of Swedish Home Sloyd: Anna P. Berg 1925
- The Teacher's Hand-Book Of Slojd: Otto Salomon 1892
- Mechanick Exercises: Joseph Moxon 1703
- The Mechanic's Companion: Peter Nicholson 1850
- The Student's Instructor In Drawing And Working The Five Orders Of Architecture: Peter Nicholson 1815
- The Circle Of The Mechanical Arts: Thomas Martin 1813
- The Complete Cabinet-Maker's And Upholsterer's Guide: J. Stokes 1829
- A Manual Of Wood Carving: Charles G. Leland, Revised by John J. Holtzapffel 1891
- Wood Carving: Joseph Phillips 1896
- Woodwork Tools And How To Use Them: William Fairham 1922
- Woodwork Joints: William Fairham 1920
- Cabinet Construction: J. C. S. Brough 1930
- Furniture Making: Advanced Projects In Woodwork: Ira Griffith 1912
- The Painter, Gilder, And Varnisher's Companion: H. C. Baird 1850
- Our Workshop: Temple Thorold 1866
- Carpentry And Joinery For Amateurs: James Lukin 1879
- The Art Of Mitring: Owen Maginnis 1892
- Working Drawings Of Colonial Furniture: F. Bryant 1922

CARPENTRY MADE EASY;

OR,

The Science and Art of Framing,

ON A NEW AND IMPROVED SYSTEM.

WITH SPECIFIC INSTRUCTIONS FOR

BUILDING BALLOON FRAMES, BARN FRAMES, MILL FRAMES, WAREHOUSES, CHURCH SPIRES, ETC.

COMPRISING ALSO

A SYSTEM OF BRIDGE BUILDING;

WITH

BILLS, ESTIMATES OF COST, AND VALUABLE TABLES.

ILLUSTRATED BY

Thirty-eight Plates and near Two Hundred Figures.

BY WILLIAM E. BELL,
ARCHITECT AND PRACTICAL BUILDER.

PHILADELPHIA:
HOWARD CHALLEN.

Entered according to Act of Congress, in the year 1857, by
WILLIAM E. BELL,
In the Clerk's Office of the District Court of the United States, in and for the Eastern District of Pennsylvania.

STEREOTYPED BY GEORGE CHARLES,
No. 607 Sansom Street, Philadelphia.

PREFACE.

The Author takes great pleasure in acknowledging the eminent services rendered him in the literary and scientific portions of this work, by E. N. JENCKS, A. M., Professor of Mathematics and Natural Sciences; and the Public cannot fail to appreciate the value of his labors in these departments.

The inception of the work, its original designs, and the entire system, are mine. Whatever is found in it purely literary and scientific, I cheerfully attribute to his assistance. And believing that the work will supply a pressing want, and will be useful both to those who are devoted to the Mechanic Arts and to Amateurs who have felt the necessity of a faithful guide in house-building and other structures, especially in new settlements, I can confidently commend it to them as supplying this deficiency.

WILLIAM E. BELL.

Ottawa, Ill., Jan. 1st, 1858.

CONTENTS

PART I.—GEOMETRY.

	PAGE
Definitions	17
Explanations of Mathematical Symbols	21
Definitions of Mathematical Terms	22
Axioms	22
Proposition I. Theorem	23
Proposition XXX. Problem	39

PART II.—CARPENTRY

USE OF THE SQUARE IN OBTAINING BEVELS	43
The square described	43
Pitch of the roof	43
Bevels of Rafters	44
Bevels of upper joints and gable-end studding	45
Bevels of Braces	45
BALLOON FRAMES	47
The sills (light sills)	47
The studs	47
The plates	47
Raising and plumbing the frame	47
The floor joists	48
Upper joists	48
Rafters	48
Gable-end studs	49
Framing the sills	50
Work sides (of timbers)	50
To take timber out of wind	50
Spacing for windows and doors	51
Mortices for the studs	51
The gains (for joints)	51
The draw bores	51
A draw pin	51
Supports for the upper joists	52
Crowning of joists	53
Bridging of joists	53
Lining, or sheeting balloon frames	53
BARN FRAMES	55
Size of mortices	55
Braces	56
Pitch of the roof	56
Purlins	56
Length of the purlin posts	57
Purlin post brace	58
Purlin post brace mortices	58
Upper end bevel of purlin post braces	60
MILL FRAMES	61
Cripple studs	61
Trussed partitions	63

CONTENTS.

	PAGE
SCARFING	64
Straps and bolts (in scarfing)	64
Scarfing over posts	65
FLOORS IN BRICK BUILDINGS	66
Trimmer joists	66
CIRCULAR CENTRES	67
ELLIPTICAL CENTRES	68
ARCHES	69
HIP ROOFS	70
Hip Rafters	70
Side bevel of hip rafters	70
Down bevel of hip rafters	71
Backing of hip rafters	71
Lengths and bevels of the jack rafters	72
HIPS AND VALLEYS	73
TRAPEZOIDAL HIP ROOFS	73
Lengths of the irregular hip rafters	73
Bevels of the irregular hip rafters	74
Backing of hip rafters on trapezoidal and other irregular roofs	75
Length of jack rafters	75
Side bevels of jack rafters on the sides of the frame	76
Side bevels of the jack rafters on the slant end of the frame	76
Down bevel of the jack rafters on the beveled end of the frame	77
OCTAGONAL AND HEXAGONAL ROOFS	78
Length of the hip rafters	78
Bevels of the hip rafters	78
Backing of the octagonal hip rafters	79
Length of the jack rafters	79
Width of the Building	79
ROOFS OF BRICK AND STONE BUILDINGS	81
Lengths and bevels of the braces	81
Dimensions of timbers for figs. 1 and 2	82
Length of straining beam	83
CHURCH SPIRES	89
DOMES	90

PART III.—BRIDGE BUILDING.

STRAINING BEAM BRIDGES	93
TRESTLE BRIDGES	98
ARCH TRUSS BRIDGES	102
GENERAL PRINCIPLES OF BRIDGE BUILDING	107

PART IV.—EXPLANATION OF THE TABLES.

Definitions of Terms and Phrases used in this Work	113
Table I. Length of Common Rafters	116
Table II. Length of Hip Rafters	119
Table III. Octagonal Roofs	125
Table IV. Length of Braces	127
Table V. Weight of Square Iron	128
Table VI. Weight of Flat Iron	130
Table VII. Weight of Round Iron	132
Table VIII. Weight and Strength of Timber	134

INTRODUCTORY CHAPTER.

SUMMARY VIEW.

The Science and the Art of Framing.

No apology is offered for introducing to the Public a work on the Science and Art of Framing. By the *Science of Framing* is meant the *certain knowledge* of it, founded on mathematical principles, and for which the master of it can assign intelligent reasons, which he knows to be correct; while the *Art of Framing* is the system of rules serving to facilitate the *practice* of it, but the *reasons* for which the workman may or may not understand. That Carpentry has its rules of Science as well as its rules of Art, no intelligent mechanic can doubt. The rules of the Art are taught by the master-workman at the bench; or, more commonly, insensibly acquired by habit and imitation. But by whom have the rules of the *Science* been laid down, and where have its principles been intelligibly demonstrated?

Something New.

It is believed that this is the very first attempt ever made to bring the Science of Carpentry, properly so called, within the scope of practical mechanics.

Deficiencies of Former Works on Carpentry.

Whatever has formerly been published on this subject, that can, with any degree of propriety, be classed under the head of Science, has been only available by professional Architects and Designers, being written in technical language and mathematical signs, accompanied by no adequate definitions or explanations; and are as perfectly unintelligible to working-men of ordinary education as Chinese or Choctaw. On the other hand, the numerous works upon the Art of Carpentry, designed and published for the use of working-men, are sadly deficient in details and practical rules. They seem to take it for granted that the student is already familiar with his business; they furnish him with drafts and plans to work from; they tell him authoritatively that such or such an angle is the proper bevel for such a part of the frame; but they neither tell him *why* it is so, nor inform him *how* to begin and go on systematically with framing and erecting a building. These works are, in fine, chiefly valuable for their plates; and even these it is not always possible to work from with confidence and accuracy, because no man can work with confidence and accuracy in the dark: and he certainly is in the dark who does not understand the reasons on which his rules are founded.

The Author's Experience.

These facts and reflections have been impressing themselves upon the mind of the Author of this work for twenty years past, while he has been serving the Public as a practical carpenter. During much of this time it has been his fortune to have large jobs on hand, employing many journeymen

INTRODUCTORY CHAPTER.

mechanics, who claimed to understand their trade, and demanded full wages. But it has been one of the most serious and oppressive of his cares, that these journeymen knew so little of their business.

Few Good Carpenters.

They had, by habit, acquired the use of tools, and could perform a job of work after it had been laid out for them; but not more than one man in ten could himself lay out a frame readily and correctly.

Why Apprentices do not Learn.

Now, it is not commonly because apprentices are unwilling to learn, or incapable of learning, that this is so, but it is because they have not the adequate instruction to enable them to become master-workmen. Their masters are very naturally desirous to appropriate their services to their own best advantage; and that is often apparently gained by keeping the apprentice constantly at one branch of his business, in which he soon becomes a good hand, and is taught but little else; and when his time is his own, and he comes to set up business for himself, then he is made to feel his deficiencies. Should he have assistants and apprentices in his turn, he would be unable to give them proper instruction, even were he well disposed to do so—for he can teach them nothing more than what he knows himself.

In this condition, the young mechanic applies to books to assist him to conquer the mysteries of his Art; but he has not been able hitherto to find a work adapted to his wants. He anxiously turns the pages of ponderous quarto and folio volumes; he is convinced of the prodigious learning of the authors, but he is not instructed by them. On the one hand, their practical directions and rules are too meagre; and, on the other hand, their mathematical reasoning is too technical to yield our young working-man any real benefit or satisfaction. May not these faults be remedied? Is it not possible for instruction to be given, which shall be at once simple and practical in detail, and comprehensible and demonstrative in mathematical reasoning?

Design of this Work.

An attempt has been made, in this little work, to answer these questions affirmatively; and thus to supply a positive want, and to occupy a new field in the literature of Architecture. Its design is to give plain and practical rules for attaining a rapid proficiency in the Art of Carpentry; and also to prove the correctness of these rules by mathematical science.

Importance of Geometry to Carpenters.

No certain and satisfactory knowledge of framing can be gained without a previous acquaintance with the primary elements of Arithmetic and Geometry. It is presumed that a sufficient knowledge of Arithmetic is possessed by most mechanics in this country; but Geometry is not so commonly understood. It is not taught in our District Schools, and is looked upon as beyond the capacity of common minds. But this is a mistake. To mechanical minds, at least, the elements of Plane Geometry are so easily taught, that they seem to them to be almost self-evident at the first careful perusal; and mechanics have deprived themselves of much pleasure, as well as profit, in not having made themselves masters of this science.

Geometry in this Work.

Part I. is therefore devoted to so much of the Science of Geometry as is essential to the complete demonstration and thorough understanding of the Science and Art of Carpentry; and it is recommended to all mechanics into whose hands this volume may fall, to give their days and nights to a careful study of this part of the work. It is true that our rules and instructions in Carpentry are so plain and minute, that they are available to those who do not care to study Geometry at all; but the principles on which those rules are founded, and consequently the *reasons why the rules are as they are*, cannot, from their very nature, be made plain and intelligible to any one except by a course of geometrical reasoning.

New Rules of Carpentry.

Part II. comprises the main body of the work, and is devoted particularly to the framing of buildings. The rules for obtaining the bevels of rafters, joists, braces, &c., as explained in this part of the work, it is believed, have never been published before. That such bevels could be so found has been known, for several years past, among master-builders; and, to a limited extent, has by that means been made public; but this feature of the work will, no doubt, be new and useful to some mechanics who have followed the business for years, and will be especially useful to apprentices and young journeymen who have not yet completed their mechanical education.

They are Proved and Explained.

These rules have been here demonstrated by a new and rigid course of geometrical reasoning; so that their correctness is placed beyond doubt. The demonstrations are often given in foot-notes and in smaller print, so as not to interrupt the descriptive portion of the work, nor appall those who are not mechanically learned, by an imposing display of scientific signs and technical terms. In fact, it has been made a leading object, in the preparation of this work, to convey correct mechanical and scientific principles in simple language, stripped as much as possible of all technicalities, and adapted to the comprehension of plain working-men.

Bridge Building.

Part III. comprises a brief practical treatise on the framing and construction of Bridges, with bills of timber and iron given in detail, by the use of which intelligent carpenters can construct almost any kind of a bridge. This part of the work does not, however, make any special claims to new discoveries, or to much originality; nor is it intended to supercede the use of those works specially devoted to Bridge Building; but it is believed it will be found more practically convenient and simple than some others of more imposing bulk and of higher price.

Valuable Tables.

Part IV. contains a valuable collection of Tables, showing the Lengths of Rafters, Hip Rafters, Braces, &c., and also the weights of iron, the strength of timber, &c., &c., which will be found of the greatest convenience, not only to common mechanics but to professional designers, architects, and bridge builders. Some of these tables have been compiled from reliable sources; but the most important of them have been calculated and constructed, at a considerable amount of expense and labor, expressly for this work.

Plates and Illustrations.

Nor has any expense been spared in the preparation of the plates and illustrations, which are "*got up*" in the highest style of the art; and it is hoped, and confidently expected, that the work, as a whole, will prove to be satisfactory and remunerative equally to the Public and to their

Humble and obedient servant,

THE AUTHOR.

Ottawa, Ill.

PART I.

Plate 1.

GEOMETRY.

PLATES I. AND II.

Definitions.

1. *Mathematics* is the science of quantity.
2. *Quantity* is any thing which can be measured, increased or diminished.
3. *The fundamental Branches* of Mathematics are Arithmetic and Geometry.*
4. *Arithmetic* is the science of numbers.
5. *Geometry* is the science of magnitude.
6. *Magnitude* has three dimensions: length, breadth, and thickness.
7. A *line* has length without thickness. The extremities of a line are called *points*. A point has no magnitude, but position only.
8. A *straight line* is the shortest distance between two points.
9. A *curved line* is one which changes its direction at every point. It is neither straight nor composed of straight lines.

Thus in Fig. 1, AB is a straight line. ACDB is a *broken* line, or one composed of straight lines; and AFB is a curved line.

10. The single term *line* is often used in the sense of *straight line;* and the single term *curve,* of *curved line.*
11. Two lines are *parallel* when they are everywhere equally distant. Fig. 2.
12. A *surface* has length and breadth without height or thickness.
13. A *plane* is a surface, in which, if any two of its points be joined by a straight line, that line will lie wholly on the surface.
14. A *solid,* or *body,* is that which combines the three dimensions of magnitude, having length, breadth, and thickness.
15. When two straight lines meet each other, the inclination or opening

---

* Algebra is a branch of Mathematics, but can scarcely be regarded as equally fundamental with Arithmetic and Geometry.

between them is called an *angle;* and this angle is said to be greater or less as the lines are more or less opened or inclined.

The *vertex* of an angle is the point where its sides meet. Thus, in Fig 3, A is the vertex, and AB and AC are the *sides*.

Angles occupy surfaces; they are therefore quantities; and like all other quantities are susceptible of addition, subtraction, multiplication, and division. Thus, in Fig. 4, the angle DCE is the sum of the two angles DCB and BCE And the angle DCB is the difference of the two angles BCE and DCE.

An angle is designated by the letter at the vertex, when there is but one angle there, as the angle A in Fig. 3; or otherwise by the three letters BAC or CAB, the letter at the vertex being always placed in the middle.

16. When a line, AB, stands on another line, CD, Fig. 5, so as not to incline either way, AB is said to be *perpendicular* to CD, and the angle on each side of the perpendicular is called a *right angle*.

17. Every angle less than a right angle is called an *acute angle*, as DCB, in Fig. 4; and every angle greater than a right angle, as ACD, is called an *obtuse angle*.

18. A *polygon* is a portion of a plane terminated on all sides by straight lines.

19. An *equilateral* polygon has all its sides equal, and an *equiangular* polygon has all its angles equal.

20. A *regular* polygon is one which is both equilateral and equiangular.

21. The polygon of three sides is called a *triangle;* that of four sides, a *quadrilateral;* one of five sides, a *pentagon;* one of six, a *hexagon;* one of seven, a *heptagon;* one of eight, an *octagon;* one of nine, a *nonagon;* one of ten, a *decagon;* one of twelve, a *dodecagon;* one of fifteen, a *pentedecagon;* and so on, according to the numerals of the Greek language.

22. An *equilateral triangle* has its three sides equal: Fig. 6. An *isosceles triangle* has two of its sides equal. A *scalene triangle* has all its sides unequal. Figs. 7 and 8.

23. A *right-angled triangle* contains one right angle. The side opposite the right angle is called the *hypotenuse*. Fig. 9: AC, opposite the right angle B, is the hypotenuse.

24. Quadrilaterals are designated according to their figures, as follows:

The *square* has its sides all equal, and its angles all right angles. Fig. 10.

The *rectangle,* or oblong square, Fig. 11, has all its angles right angles and its opposite sides equal and parallel.

The *parallelogram,* Fig. 12, has its opposite sides equal and parallel. Every rectangle is a parallelogram, but every parallelogram is not a rectangle.

The *rhombus,* or *lozenge,* has its sides all equal without having its angles right angles. Fig. 13.

The *trapezium* has none of its sides parallel. Fig. 14.

The *trapezoid* has two of its sides parallel. Fig. 15.

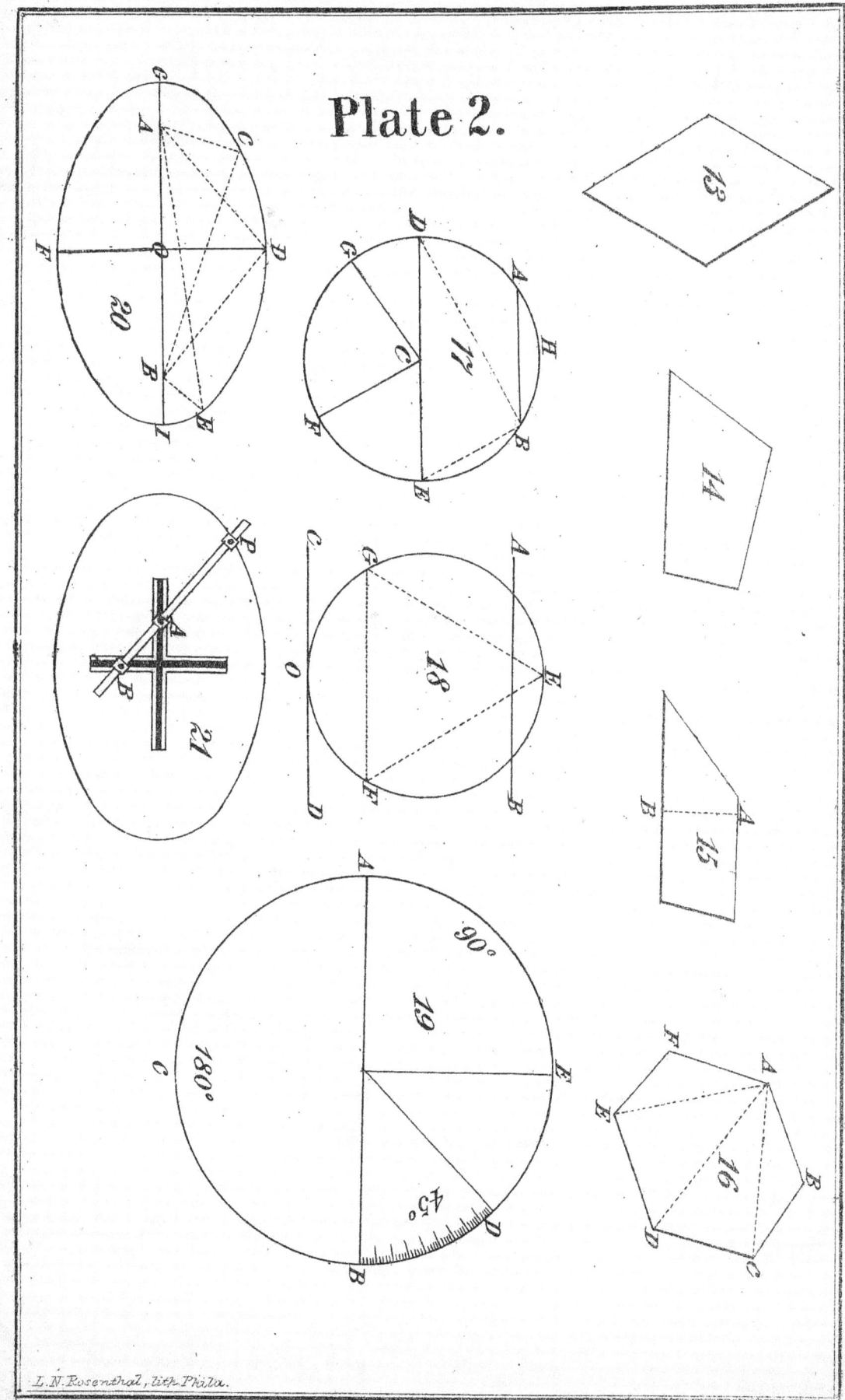

25. The *base* of any polygon is the side on which it is supposed to stand.

26. The *altitude of a triangle* is the perpendicular let fall upon the base from the vertex of the angle opposite the base. Thus, in Fig. 6, AB is the altitude of the triangle ACD.

27. The *altitude of a parallelogram*, or of a trapezoid, is the perpendicular which measures the distance between two parallel sides taken as bases. Thus, in Fig. 12, AB is the altitude of the parallelogram CD.

28. A *diagonal* is a line within a polygon, which joins the vertices of two angles not adjacent to each other. Thus, in Fig. 16, AC, AD, and AE are diagonals.

29. The *area* of a polygon is the measure of its surface.

30. *Equivalent polygons* are those which contain equal areas.

31. *Equal polygons* are those which coincide with each other in all their parts. (Ax. 13.)

32. *Similar polygons* have the angles of the one equal to the angles of the other, each to each, and the sides about the equal angles proportional.

33. *Homologous sides* and *homologous angles* are those which have like positions in similar polygons.

34. The *circumference* of a circle is a curved *line*, every point of which is equally distant from a point within, called the *centre*.

35. The *circle* is the *surface* bounded by the circumference.

36. A *radius* of a circle is a straight line drawn from the centre to the circumference. (The term *radius* is a Latin word, the plural of which is *radii*. Thus, we say one radius and two radii.) In the same circle all radii are equal; all diameters are also equal, and each diameter is double the radius.

37. A *diameter* of a circle is a straight line drawn through the centre, and terminated on both sides by the circumference. In Fig. 17, CD, CG, and CF are radii, and DE is a diameter.

38. An *arc* is a portion of the circumference; as AHB in Fig. 17.

39. A *chord* is the straight line which connects the two extremities of an arc. AB, Fig. 17.

40. A *segment* is a portion of a circle included between an arc and its chord; as segment AHB, Fig. 17.

41. A *sector* is a portion of the circle included between two radii; as sector CGF, Fig. 17.

42. An *inscribed angle* is one formed by the intersection of two chords upon the circumference. ABD and BDE are inscribed angles.

43. An *inscribed polygon* is one which, like EFG in Fig. 18, has all its angles in the circumference. The circle is then said to circumscribe such a figure. In Fig. 17, the triangle CGF is *not* an inscribed triangle, since all the angles do not lie in the circumference; but, in the same Fig., DBE *is* an inscribed triangle.

44. A *secant* is a line which intersects the circumference in two points, and lies partly within and partly without the circle. AB is a secant in Fig. 17.

45. A *tangent* is a straight line touching the circumference in one point only. CD is a tangent—O is the point of contact.

46. The circumference of every circle is measured by being supposed to be divided into 360 equal parts, called *degrees;* each degree contains 60 *minutes*, and each minute 60 *seconds*. Degrees, minutes, and seconds, are designated respectively by these characters, °, ', " ; thus, 45° 15' 30" is read 45 degrees, 15 minutes, and 30 seconds.

47. *Arcs are measured* by the number of degrees which they contain. Thus, in Fig. 19, the arc AE, which contains 90 degrees, is called a *quadrant*, or the quarter of a circumference, because 90° is one quarter of 360°; and the arc ACB, which contains 180°, is a semicircumference.

48. *Every angle is also measured by degrees;* these degrees being reckoned on an arc included between its sides, described from the vertex of the angle as a centre. Thus, in Fig. 19, the right angle AOE contains 90°; and the angle BOD, which is one half a right angle, is called an angle of 45 degrees, which is the number it contains.

49. An *ellipse* is a curved line drawn around two points within, called *foci** (A, B, Fig. 20), in such a manner that if, from any point, C, of the curve, two lines be drawn, one to each focus, the sum of these two lines, AC and BC, shall be equal to the sum of two other lines drawn to the foci from any other point of the curve; as DA and DB, or EA and EB.

50. The *centre of an ellipse* is the middle point of the line joining the two foci. O in Fig. 20.

51. The *diameter of an ellipse* is any straight line passing through the centre, and terminated on both sides by the curve.

52. The *conjugate axis* of an ellipse is its longest diameter, or that one which passes through the two foci; as GI.

53. The *transverse axis* of an ellipse is its shortest diameter, or that one which is perpendicular to the conjugate axis; as DF.

Note. There are several methods employed *to describe an ellipse*, but the one which is at once the most correct and practicable is by means of the instrument called the *Trammel*, represented in Fig. 21. It consists of two grooved rules, mitered together in the middle, so that each arm is perfectly perpendicular to the two adjacent arms, and a rod with a movable pencil at P, and two movable pins, one at A and the other at B.

The distance from B to P equals half the conjugate axis.

The distance from A to P equals half the transverse axis.

The distance from A to B equals half the distance of either focus from the centre.

---

* The term *focus* is a Latin word, of which the plural is foci; thus, we say one *focus* and two *foci*.

Explanation of Mathematical Symbols.

In order to facilitate mathematical calculations, it has long been customary among civilized nations not only to employ figures to represent numbers, but also to employ certain other signs or symbols to represent such operations, in the combinations of numbers and quantities, as are of most frequent occurrence; and, by mutual consent, these symbols have come to be generally known and employed for this purpose.

1. The sign of *addition* is written thus $+$, and is read *plus;* for example, $2+3$ is read two plus three, and signifies two added to three.

2. The sign of *subtraction* is written thus $-$, and is read *minus;** for example, $3-2$ is read three minus two, and signifies three less two.

3. The sign of *multiplication* is written thus $\times$, and is read *multiplied by;* thus, 3×2 is read three multiplied by two.

4. The sign of *division* is written thus $\div$, and is read *divided by;* thus, $12\div 3$ signifies 12 divided by 3. Division is more commonly indicated, however, by writing the divisor under the dividend, with a line between them, in the form of a fraction; thus, $\frac{12}{3}$ signifies, as before, 12 divided by three.

5. The sign of *equality* is written thus $=$, and is read *equals,* or *is equal to;* for example, $2+3=5$ is read thus, two plus three equals five.

6. The letters of the alphabet, A, B, C, &c., are used as *representatives* of quantities, the exact dimensions of which may either be known or unknown. We can let A, for example, stand for a given line, a given angle, a given square, or a given solid. Lines are most commonly represented by the two letters placed at their extremities; and angles by their three letters, the letter at the vertex being always placed in the middle.

7. A number placed before a quantity is called a *co-efficient;* thus, $5AB$ is read five AB, or five times AB, or AB multiplied by five: the sign of multiplication being understood but not written.

8. A number placed at the right, and a little above a quantity, is called an *exponent*, and indicates how many times a quantity is taken as a factor; thus, 5^2 is read *five square;* it is equal to 5×5, and signifies that five is to be multiplied by five, which equals 25; also, 5^3 is read *five cube;* it is equal to $5\times 5\times 5$, and signifies that five is to be multiplied by five, and that product by five, which equals 125.

9. This sign $\sqrt{\ }$ is used to show that a *root is to be extracted.* A small figure is placed in the bosom of the sign, called the *index of the root;* thus, $\sqrt[2]{\ }$ is the sign of the square root, and $\sqrt[3]{\ }$ is the sign of the cube root, &c. When no index is written, that of the square root is understood; thus, $\sqrt{4}$ represents the square root of 4.

---

* *Plus* and *minus* are Latin words, the former meaning *more*, and the latter *less;* these words, like the signs, are in common use in all civilized countries.

Definitions of Mathematical Terms.

1. An *axiom* is a self-evident truth.

2. A *theorem* is a statement which requires a demonstration, by reasoning from such truths as are either self-evident or previously demonstrated.

3. A *problem* is a query to be answered, or an operation to be performed.

4. The term *proposition* may be applied either to axioms, theorems, or problems.

5. A *corollary* is a necessary inference drawn from one or more preceding propositions.

6. A *scholium* is an explanatory remark on one or more preceding propositions.

7. An *hypothesis* is a supposition employed either in the statement or the demonstration of a proposition.

8. The term *ratio* is employed to denote the quotient arising from dividing one number or quantity by another: for example, the ratio of 3 to 12 is 12 divided by 3; or; $\frac{12}{3}$ or 4. The ratio can always be expressed in the form of a fraction, whether the divisor is contained in the dividend an exact number of times or not; thus the ratio of 2 to 1 is $\frac{1}{2}$, the ratio of 5 to 6 is $\frac{6}{5}$; and so also the ratio of A to B is $\frac{B}{A}$, and the ratio of x to y is $\frac{y}{x}$.

9. *Proportion* is an equality of ratios or an *equality of quotients*. Thus when the quotient arising from dividing one quantity by another is equal to the quotient arising from dividing a third quantity by a fourth, then the four quantities are said to be in proportion to each other. For example, the quotient of 4 divided by 2 equals the quotient of 10 divided by 5; or $\frac{4}{2} = \frac{10}{5}$; then these four numbers 2, 5, 4 and 10 are in proportion.

Proportion is usually indicated by writing the four quantities thus: 2 : 4 : : 5 : 10, and is read 2 is to 4 as 5 is to 10; that is, 2 is just such a part of 4 as 5 is of 10; for 2 is half of 4, and 5 is half of 10. So also if $\frac{B}{A} = \frac{D}{C}$, then we have the proportion A : B : : C : D.

10. The four quantities of a proportion are called its *terms*. The first and last are called the *extremes*, and the two middle ones the *means* of a proportion. The first and third terms are called the *antecedents*, and the second and fourth terms are called the *consequents* of a proportion.

Axioms.

1. A whole quantity is greater than any of its parts.
2. A whole quantity is equal to the sum of all its parts.
3. When equals are added to equals, their sums are equal.
4. When equals are added to unequals, their sums are unequal.
5. When equals are subtracted from equals, their remainders are equal

GEOMETRY. 23

6. When equals are subtracted from unequals, their remainders are unequal.

7. When equals are multiplied by equals, their products are equal.

8. When equals are divided by equals, their quotients are equal.

9. When two quantities have, each, the same proportion to a third quantity they are equal to each other.

10. All right angles are equal.

11. When a straight line is perpendicular to one of two parallels it is perpendicular to the other also.

12. Only one straight line can be drawn from one point to another.

13. Two magnitudes are equal, when, on being applied to each other, they coincide throughout their whole extent.

Proposition I. Theorem.

If four quantities are in proportion, the product of the two means will equal the product of the two extremes.

	Numerically.	Generally.
Let	$2:4::5:10;$	$A:B::C:D;$
then will	$4 \times 5 = 2 \times 10.$	$B \times C = A \times D.$

For, since the given quantities are in proportion, their ratios are equal (Def. of Terms, 9.)

And we have, $\quad \dfrac{4}{2} = \dfrac{10}{5}. \qquad\qquad \dfrac{B}{A} = \dfrac{D}{C}.$

Multiply both quantities by the divisor of the first ratio, and the quantities will still be equal (Ax. 7); we shall then have,

$$2 \times \frac{4}{2} = 2 \times \frac{10}{5}; \qquad\qquad A \times \frac{B}{A} = A \times \frac{D}{C};$$

or, $\qquad 4 = 2 \times \dfrac{10}{5}. \qquad\qquad B = A \times \dfrac{D}{C}.$

Again, multiply both quantities by the divisor of the second ratio, and the desired result is obtained; namely,

$$4 \times 5 = 2 \times 10. \qquad\qquad B \times C = A \times D.$$

Proposition II. Theorem.

When the product of two quantities equals the product of two other quantities, then two of them are the means, and the other two the extremes of a proportion.

	Numerically.	Generally.
Let	$4 \times 5 = 2 \times 10;$	$B \times C = A \times D;$
then will	$2:4::5:10;$	$A:B::C:D;$

for, divide both the given quantities by one of the factors of the first quantity, which will not alter their equality (Ax. 8), and we have,

$$4 = \frac{2 \times 10}{5}; \qquad\qquad B = \frac{A \times D}{C};$$

again, divide both quantities by one of the factors of the second quantity, and we have,

$$\frac{4}{2}=\frac{10}{5}. \qquad \frac{B}{A}=\frac{D}{C}.$$

Here we have an equality of ratios, and, by Def. 9, the four quantities are in proportion; hence,

$$2:4::5:10. \qquad A:B::C:D.$$

Scholium. Quantities are said to be in proportion by *inversion*, when the proportion is read backward;

thus, $4:2::10:5;$ $B:A::D:C;$
or, $10:5::4:2.$ $D:C::B:A.$

Quantities are said to be in proportion by *alternation*, when they are read alternately;

thus, $2:5::4:10;$ $A:C::B:D;$
or, $4:10::2:5.$ $B:D::A:C.$

Quantities are said to be in proportion by *composition*, when the sum of the antecedents or consequents is compared with either antecedent or consequent.

thus, $2+5:5::4+10:10;$ $A+B:B::C+D:D;$
or, $2+5:2::4+10:4.$ $A+B:A::C+D:C.$

Proposition III. Theorem.

When four quantities are in proportion, they will also be in proportion by alternation.

	Numerically.	Generally.
Let	$2:4::5:10,$	$A:B::C:D,$
then will	$2:5::4:10;$	$A:C::B:D;$
for, by Prop. I.,	$2\times10=5\times4;$	$A\times D=C\times B;$
and, by Prop. II.,	$2:5::4:10.$	$A:C::B:D.$

Proposition IV. Theorem.

When four quantities are in proportion, they will also be in proportion by inversion.

	Numerically.	Generally.
Let	$2:4::5:10,$	$A:B::C:D,$
then will	$10:5::4:2;$	$D:C::B:A;$
for, by Prop. I.,	$10\times2=5\times4;$	$D\times A=C\times B;$
and, by Prop. II.,	$10:5::4:2.$	$D:C::B:A.$

Proposition V. Theorem.

When there are four proportional quantities, and four other proportional quantities, having the antecedents the same in both, the consequents will be proportional.

	Numerically.	Generally.
Let	$2:4::5:10$, and	$A:B::C:D$,
and	$2:6::5:15$;	$A:X::C:Y$;
then will	$4:10::6:15$,	$B:D::X:Y$.

Take the first proportion by alternation:

$$2:5::4:10; \qquad A:C::B:D;$$

hence, from equality of ratios (Def.),

$$\frac{5}{2}=\frac{10}{4}. \qquad \frac{C}{A}=\frac{D}{B}.$$

Take the second proportion by alternation:

$$2:5::6:15, \qquad A:C::X:Y;$$

and, by equality of ratios, we have,

$$\frac{5}{2}=\frac{15}{6}; \qquad \frac{C}{A}=\frac{Y}{X};$$

hence,

$$\frac{10}{4}=\frac{15}{6}; \qquad \frac{D}{B}=\frac{Y}{X};$$

and from this equality of ratios there results (by Def.),

$$4:10::6:15. \qquad B:D::X:Y$$

Corollary. When there are two sets of proportional quantities, having an antecedent and a consequent of the first equal to an antecedent and a consequent of the second, the remaining quantities are proportional.

Proposition VI. Theorem.

When four quantities are in proportion, they are also in proportion by composition.

	Numerically.	Generally.
Let	$2:4::5:10$,	$A:B::C:D$,
then will	$2+4:2::5+10:5.$	$A+B:A::C+D:C.$

The first proportion gives (Prop. I.),

$$2\times 10 = 4\times 5. \qquad A\times D = B\times C.$$

Add to each of these equal quantities the product of the two antecedents, and we have,

$$2\times 10+2\times 5=4\times 5+2\times 5; \qquad A\times D+A\times C=B\times C+A\times C;$$

or, the same simplified,

$$2\times 10+5=5\times 4+2; \qquad A\times D+C=C\times B+A;$$

hence, by Prop. II.,

$$2+4:2::5+10:5. \qquad A+B:A::C+D:C.$$

Proposition VII. Theorem.

If any two quantities be each multiplied by some other quantity, their products will have the same ratio as the quantities themselves.

	Numerically.	Generally.
Let	2 and 4	A and B
be any two numbers;		be any two quantities;
multiply each by	5;	S;
then	$2 \times 5 : 4 \times 5 :: 2 : 4$;	$A \times S : B \times S :: A : B$;
for,	$(2 \times 5) \times 4 = (4 \times 5) \times 2$,	$(A \times S) \times B = (B \times S) \times A$,
since the quantities are identical;		
hence, by Prop. II.,	$2 \times 5 : 4 \times 5 :: 2 : 4$.	$A \times S : B \times S :: A : B$.

Proposition VIII. Theorem.

When two triangles have two sides and the included angle of the one equal to two sides and the included angle of the other, each to each, the two triangles are equal.

In the triangles ABC and DEF, let AB=DE, AC=DF, and the angle A= angle D; the triangles themselves will then be equal. For, apply the side AB to the equal side DE, so that the point A will fall upon D, and the point B upon E; then since angle A= angle D, the side AC will also fall upon its equal side DF, and the point C upon the point F; therefore the third side, CB, will fall upon the third side FE, and the two triangles will coincide throughout their whole extent, and be therefore equal. (Ax. 13.)

Proposition IX. Theorem.

When two triangles have two angles and the included side of the one, equal to two angles and the included side of the other, each to each, the two triangles are equal.

In the triangles ABC and DEF, let the angle A= angle D, C = F, and the included side AC = DF; then are the triangles also equal.

For, apply the side AC to its equal side DF, placing the point A upon the point D, and the point C upon F; then, since the angle A= angle D, the side AB will take the direction of DE, and the point B will fall somewhere upon the line DE; also, since the angle C=angle F, the side CB will take the direction of FE, and the point B will fall somewhere upon the line FE; and since the point B must fall upon both the lines DE and FE, it must fall upon E, the only point of coincidence; hence the two triangles coincide throughout their whole extent, and are therefore equal. (Ax. 13.)

Corollary. Every triangle has six parts, namely: three sides and three angles; and whenever two triangles are equal to each other, each of the six parts of the one are always equal to the corresponding six parts of the other, side to side, and angle to angle. It is to be observed, also, that the equal angles are always opposite to the equal sides, and the equal sides opposite the equal angles.

Proposition X. Theorem.

When a straight line meets another straight line, the sum of the two adjacent angles are equal to two right angles.

Let CD meet AB at D, then is the sum of the two angles ADC and CDB equal to two right angles.

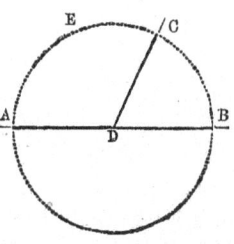

From the point D as a centre, describe the circumference of a circle, then will the line AB coincide with its diameter, since it passes through the centre. (Def.) Angles are measured by the arcs intercepted by their sides (Def.); and since the sides of the angle ADC intercept a portion of the semicircumference, AEB, and the sides of the angle CDB intercept the remaining portion, then, both together intercept a semicircumference, or 180 degrees; but two right angles intercept 180 degrees (Def.); therefore the sum of the two angles ADC and CDB=two right angles.

Cor. 1. When one of the given angles is a right angle, the other is a right angle also.

Cor. 2. When one line is perpendicular to another, then is the second line also perpendicular to the first.

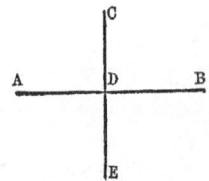

Let CE be perpendicular to AB, then is AB perpendicular to CE.

For, since CE is perpendicular to AB, both the angles ADC and CDB are right angles. Again, since AD is a straight line meeting another straight line CE at D, then the sum of ADC+ADE=two right angles; but ADC is a right angle; therefore must ADE be a right angle also. Hence AD, or AB, is perpendicular to CE.

Cor. 3. When any number of angles have their vertices at the same point, and lie on the same side of a straight line, their sum is equal to two right angles, for they all together intercept an arc of 180°

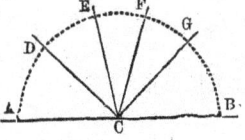

Proposition XI. Theorem.

The opposite or vertical angles, formed by the intersection of two straight lines, are equal.

Let AB and CD be two straight lines, intersecting each other at E, then will AEC=BED.

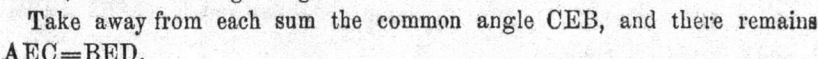

For the sum of AEC+CEB = two right angles (Prop. X.); and, for a similar reason, the sum of CEB+BED= two right angles.

Take away from each sum the common angle CEB, and there remains AEC=BED.

In a similar manner it may be proved that CEB=AED.

Proposition XII. Theorem.

If two parallel straight lines meet a third line, the sum of the two interior angles, on the same side of the line met, will be equal to two right angles.

Let the two parallel lines AB and CD meet the line EF; then will BEF+EFG=two right angles.

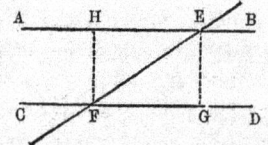

Through E draw EG, perpendicular to CD, and through F draw FH, parallel with EG. Then, since parallels are everywhere equally distant, (Def. 10), we have EH=GF, and also EG=HF; and since AB is perpendicular to EG, it is also perpendicular to HF, (Ax. 11,) and the angles H and G are both right angles; therefore, the two triangles, EHF and FGE, are equal, (Prop. VIII.) And, since the angles opposite the equal sides are equal, (Prop. IX. Cor.), angle FEH= angle EFG.

But the sum of the angles BEF+FEH is equal to two right angles. (Prop. X.) Substitute for FEH its equal EFG, and we have BEF+EFG= two right angles.

Scholium. Where two parallel straight lines meet a third line, the angles thus formed take particular names, as follows:

Interior angles on the same side are those which lie within the parallels, and on the same side of the secant line. Thus BEF and EFD are interior angles on the same side; and so also are the angles AEF and EFC.

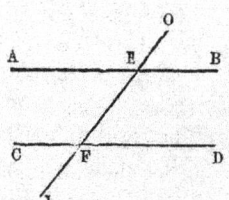

Alternate angles lie within the parallels, and on opposite sides of the secant line, but not adjacent to each other. AEF and EFD are alternate angles; also, BEF and EFC.

Alternate exterior angles lie without the parallels, and on opposite sides of the secant line. OEB and CFL are alternate exterior angles, and so also are AEO and LFD.

Opposite exterior and interior angles lie on the same side of the secant line, the one without and the other within the parallels, but not adjacent; thus OEB and EFD are opposite exterior and interior angles; so also are BEF and DFL.

Cor. 1. If a straight line meet two parallel lines, the alternate angles will be equal. For the sum BEF+EFD= two right angles; also, (by Prop. X), BEF+AEF= two right angles; take away from each the angle BEF, and there remains EFD=AEF.

Cor. 2. If a straight line meet two parallel lines, the opposite exterior and interior angles will be equal. For the sum BEF+EFD= two right angles; also, (by Prop. I.), BEF+OEB= two right angles; taking from each the angle BEF, and there remains EFD=OEB.

Cor. 3. Hence of the eight angles formed by a line cutting two parallel lines obliquely, the four acute angles are equal to each other, and so also are the four obtuse angles.

Proposition XIII. Theorem.

If two straight lines meet a third line, making the sum of the interior angles on the same side equal to two right angles, the two lines will be parallel.

Let the two lines AB, CD, meet the third line EF, so as to make the angles BEF+EFG equal to two right angles; then will AB and CD be parallel, or everywhere equally distant.

Through E draw EG perpendicular to CD, and through F draw FH parallel with EG, then the two angles FEB+FEH= two right angles, (by Prop. X.); also, the angles FEB+EFG = two right angles, by hypothesis; take away from each the angle FEB, and there remains the angle FEH= angle EFG. Again, since HF and EG are parallel by construction, the alternate angles EFH and GEF are equal, (by last Prop., *Cor.* 1); hence, the two triangles EFH and EFG are equal, (Prop. IX.), having two angles and the included side of the one equal to two angles and the included side of the other; and HF, opposite the angle FEH, is equal to EG, opposite to its equal angle EFG. (Prop. IX., *Cor.*) But HF and EG measure the distance of the line CD from the line AB, at the points H and E respectively. The same demonstration may be applied to any other two points of the line AB; hence the lines AB and CD are everywhere equally distant, and therefore parallel.

Cor. 1. If two straight lines are perpendicular to a third line, they are parallel to each other; for the two interior angles on the same side are, in that case, both right angles.

Cor. 2. If a straight line meet two other straight lines, so as to make the alternate angles equal to each other, the two lines will be parallel.

Let OL meet AB and CD, so as to make AEL= EFD; add to each the angle BEF; we shall then have AEL+BEF=EFD+BEF; but AEL+BEF= two right angles (Prop. VIII.); hence, EFD+BEF= two right angles: therefore, AB and CD are parallel.

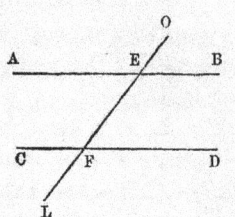

Cor. 3. If a straight line, OL, meet two other straight lines, AB and CD, so as to make the exterior angle, OEB, equal to the interior and opposite angle, EFD, the two lines will be parallel: for, to each add the angle BEF; we shall then have OEB+BEF=EFD+BEF: but OEB+BEF are equal to two right angles; therefore, EFD+BEF is equal to two right angles; and AB and CD are parallel.

Proposition XIV. Theorem.

In every parallelogram, the opposite angles are equal.

Let ABCD be a parallelogram; then will A=C, and B=D.

Draw the diagonal BD; then will the triangle ADB=the triangle CBD: for the angles ABD and BDC are alternate angles and equal (Prop. XII., *Cor.* 1), and the adjacent sides, AB=DC, and BD is common; hence, the triangles are equal (Prop. VIII.); therefore the angles A and C, opposite the common side BD, are equal. (Prop. IX., *Cor.*) In a similar manner it may be proved that the angles B and D are equal.

Cor. 1. The diagonal of a parallelogram divides it into two equal triangles.

Cor. 2. When two triangles have the three sides of the one equal to the three sides of the other, the angles opposite the equal sides are also equal, and the triangles themselves are equal.

Cor. 3. Two parallels, included between two other parallels, are equal.

Cor. 4. If the opposite sides of a quadrilateral are equal, each to each, the equal sides will also be parallel, and the figure will be a parallelogram; for, having drawn the diagonal BD, the triangles ABD and BDC are equal; and the angle ADB, opposite AB, is equal to the angle DBC, opposite DC. But the two angles, ADB and DBC are alternate angles; therefore, AD is parallel with BC. (Prop. XIII., *Cor.* 2.) ABD and BDC are also equal alternate angles; therefore, AB is parallel with DC, and the figure is a parallelogram.

Proposition XV. Theorem.

When two angles have their sides parallel, and lying in the same direction, they are equal.

Let ABC and DEF be two angles, having the side AB, in one, parallel to DE, in the other, and BC parallel to EF, and lying in the same direction; then will the two given angles be equal.

For, produce the side AB till it intersects EF at G, then ABC=BGF, for they are opposite interior and exterior angles (Prop., XII. Cor. 2); also, DEF and BGF are equal for a similar reason: therefore, ABC and DEF, being each equal to BGF, are equal to each other. (Ax. 9.)

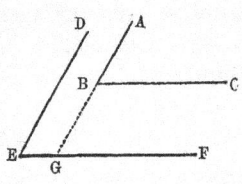

Proposition XVI. Theorem.

The sum of the three angles of any triangle is equal to two right angles.

Let ABC be any triangle. Produce the base AB to any convenient distance, as D, and draw BE parallel with AC; then will the three angles having their vertices at B be equal to the three angles of the given triangle, for the angle B, or ABC, is common; the angle c=C, for they are alternate angles; and the angle a=A, for they are opposite exterior and interior angles. But the sum of the three angles at B are equal to two right angles (Prop. X., *Cor.* 3); hence the sum of the three angles of the given triangle, A+B+C, is equal to two right angles.

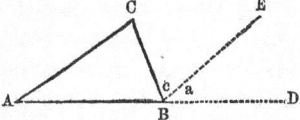

Cor. 1. The exterior angle, CBD, of any triangle formed by producing the base, is equal to the sum of the two opposite interior angles of the triangle.

Cor. 2. When the sum of two angles of any triangle is known, the third angle is found by subtracting that sum from two right angles or 180°.

Cor. 3. When two angles of one triangle are respectively equal to two angles of another triangle, their third angles are also equal, and the triangles are equiangular.

Cor. 4. It is impossible for any triangle to have more than one right angle, for if it could have two right angles, the third angle would be nothing. Still less can any triangle have more than one obtuse angle.

Cor. 5. In every right-angled triangle, the sum of the two acute angles is equal to one right angle.

Proposition XVII. Theorem.

In every isosceles triangle, the angles opposite the equal sides are equal.

In the triangle ABC let AC=BC; then will angle A= angle B. Draw the line CD so as to bisect the angle C, that is, so as to divide it into two equal parts; then are the two triangles ACD and BCD equal by Prop. VIII., having the two angles at C and the two adjacent sides equal. Hence, angle A= B. (Prop. IX., *Cor.*)

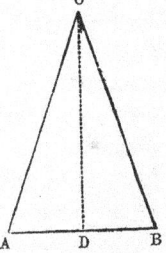

32   CARPENTRY MADE EASY.

Cor. 1. Every equilateral triangle is also equiangular.

Cor. 2. The equality of the triangles ADC, and BDC, proves that the line which bisects the vertical angle of an isosceles triangle is perpendicular to the base at its middle point, for the two angles at D are each right angles. (Prop. X.)

Proposition XVIII. Theorem.

When two angles of a triangle are equal, the sides opposite them are also equal, and the triangle is isosceles.

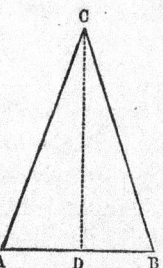

Let the angle A=B, then will the sides AC and BC be equal also.

Draw CD so as to bisect the angle C, then will the two triangles be equiangular (Prop. XVI., *Cor.* 3); and the side CD being common, the two triangles are equal (Prop. IX.); and the side AC, opposite the angle B, is equal to the side BC, opposite the equal angle A.

Proposition XIX. Theorem.

Parallelograms having equal bases and equal altitudes, contain equal areas, or are equivalent.

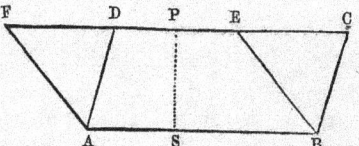

Let the two parallelograms, ABCD and ABEF, have the same base, AB, and the same altitude PS; then they will be equivalent.

In the triangles BCE and ADF, the sides BC and AD are equal, being opposite sides of the same parallelogram; and AF=BE for a similar reason; the included angle A is equal to the included angle B, since their sides are parallel and lie in the same direction (Prop. XV.); hence the two triangles are equal. (Prop. VIII.)

Now, from the whole quadrilateral figure ABCF, take away the triangle BCE, and there remains the parallelogram ABEF; from the same quadrilateral take away the equal triangle ADF, and there remains the parallelogram ABCD, which is therefore equivalent to ABEF.

Proposition XX. Theorem.

Every triangle contains half the area of a parallelogram of equal base and equal altitude.

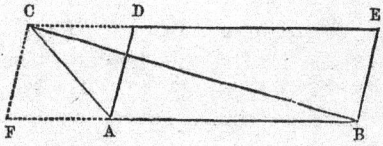

Let ABC be any triangle, and ADBE be a parallelogram having the same base and altitude; then will the triangle contain half the area of the parallelogram.

Connect C and D, and complete the parallelogram ADCF. The triangle BCF is half the parallelogram FE, (Prop. XIV., Cor. 1); and the triangle ACF is half the parallelogram FD. If from the parallelogram FE we take the parallelogram FD, then the parallelogram AE will remain; and if from the triangle BCF, half the parallelogram FE, we take the triangle ACF, half the parallelogram FD, there will remain the triangle ABC, equal to one half the parallelogram AE.

Cor. 1. The demonstrations in this and the preceding propositions, are equally applicable to rectangles, since every rectangle is also a parallelogram; therefore, every rectangle is equivalent to a parallelogram of the same base and altitude.

Also, every triangle is equivalent to half a rectangle of the same base and altitude.

Cor. 2. Triangles are equivalent to each other, when they have equal bases and equal altitudes; each being half an equivalent parallelogram.

Proposition XXI. Theorem.

Two rectangles having the same altitude are proportioned to each other as their bases.

Let the two rectangles AE and CF have equal altitudes, then will their surfaces be proportional to the length of their bases.

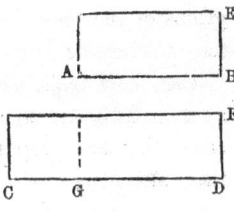

For, since their altitudes are the same, and their angles are all right angles, they may be so applied to each other that the whole surface of the shorter rectangle shall perfectly coincide with an equal surface of the longer one; and this coincidence will be perfect as far as there is a coincidence of their bases, and no further; hence,

$$AE : CF :: AB : CD.$$

Proposition XXII. Theorem.

Rectangles are proportioned to each other as the products of their bases multiplied by their altitudes.

Let P be any rectangle, having BC for its base, and BF for its altitude; and let N be any other rectangle, having AB for its base, and BE for its altitude; then,

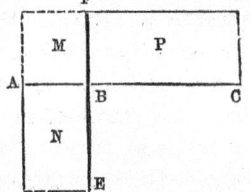

$$P : N :: BC \times BF : AB \times BE.$$

For, place the two rectangles P and N, so that the base AB will be the prolongation of the base BC, and complete the rectangle M; then, the two rectangles P and M, having the same altitude BF, will be proportioned to each other as their bases, CB and AB (Prop. XXI.) And, for the same reason, the two rectangles N and M, having the same altitude AB, will be to each other as their bases BE and BF; hence, we have the two proportions:

$$P : M :: BC : AB; \text{ and}$$
$$M : N :: BF : BE.$$

Combining these two proportions, by multiplying the corresponding terms together, we have

$$P \times M : N \times M :: BC \times BF : AB \times BE.$$

But the quantity M, since it is common to both antecedent and consequent, can be omitted; and the remaining quantities will still be proportional. (Prop. VII.) Hence,

$$P : N :: BC \times BF : AB \times BE.$$

Cor. 1. Hence the area or surface of any rectangle is measured by the product of its base multiplied by its altitude; and if its base be BC, and its altitude BF, its area or measure is $BC \times BF$.

Cor. 2. Since the sides of every square are all equal, and since all squares are rectangles, the area of any square is expressed by the product of a side multiplied by itself: so if its side is AB, its area is AB^2.

Cor. 3. Since every rectangle is a parallelogram, and since all parallelograms of the same base and altitude are equivalent, (Prop. XIX.), therefore the area of any parallelogram is the product of its base by its altitude.

Cor. 4. Parallelograms of the same base are proportioned to each other as their altitudes, and those of the same altitude as their bases; and, in all cases, they are proportioned to each other, as the products of their bases by their altitudes.

Proposition XXIII. Theorem.

The area of any triangle is measured by the product of its base multiplied by half its altitude.

Let ABC be any triangle, of which AB is the base, and CD the altitude. This triangle is half the parallelogram AE, (Prop. XIV., *Cor.*); but the parallelogram is measured by its base, AB, multiplied by its altitude, DC; therefore the triangle is measured by the base multiplied by half the altitude.

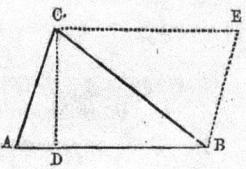

Cor. Triangles of the same altitude are proportioned to each other as their bases, and those of the same bases are to each other as their altitudes; and, in any case, they are proportioned to each other as the products of their bases by their altitudes.

Proposition XXIV. Theorem.

In every right-angled triangle, the square of the hypotenuse is equal to the sum of the squares of the other two sides.

Let ABC be a triangle, having the angle C a right angle; then will $AB^2 = AC^2 + CB^2$.

GEOMETRY. 35

Complete the squares of the three sides of the given triangle, and let M represent the square described on AB, or AB^2; let N represent the square described on CB, or CB^2; and let P represent the square described on AC, or AC^2. Draw the diagonals DB, CE, CI, and AH, and from C let fall CG, perpendicular to AB.

In the two triangles DAB and CAE, AC=AD, each being a side of the square P; and AB=AE, each being a side of the square M; the included angle DAB is made up of the right angle DAC and the angle CAB; the included angle CAE, in the other triangle, is made up of the same angle CAB, and the right angle BAE; hence, the angles CAE and DAB are equal, and the triangles themselves are equal (Prop. VIII.), each having two sides and an included angle equal.

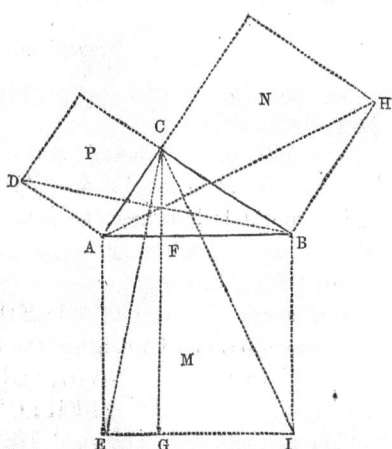

The triangle DAB is equivalent to half the square P, for it has the same base AD, and the same altitude AC; also, the triangle CAE is equivalent to half the rectangle FAGE, for it has the same base AE, and the same altitude AF; hence, the rectangle FAGE is equivalent to the square P. Again, the two triangles ABH and CBI are equal, having also two sides and the included angle of the one equal to two sides and the included angle of the other; and AHB is half of the square N, and CBI is half of the rectangle FBIG: therefore, the square N is equivalent to the rectangle FBIG.

But the two rectangles FAGE and FBIG make up the square M; hence, $M=P+N$, or $AB^2=AC^2+CB^2$.

Cor. 1. In every right-angled triangle, the square of one side is equivalent to the square of the hypotenuse less the square of the other side. For example, in the triangle above, $AC^2 = AB^2-BC^2$; also, $BC^2=AB^2-AC^2$.

Cor. 2. Every square is equal to half the square of its own diagonal.

Let AD and BC be the diagonals of the square ABCD; through the points A and D draw straight lines equal and parallel with BC; and through the points B and C draw lines equal and parallel with AD; the figure thus formed will be the square of the diagonal CB, or of its equal EF: but this figure contains eight equal triangles, of which the given square contains but four; hence,

$$CB^2 : AB^2 : : 2 : 1$$

and, on extracting the square root of each of the terms of this proportion, we have,

$$CB : AB :: \sqrt{2} : 1;$$

or, the diagonal of a square is proportioned to its side, as the square root of two is to one.

Proposition XXV. Theorem.

In any triangle, a line drawn parallel to the base divides the other two sides proportionally.

Let ABC be any triangle, and let DE be parallel with the base AB. Draw AE and BD. The two triangles ADE and CDE, having the same altitude DE, are in proportion to each other as their bases AD and CD (Prop. XXIII., *Cor.*); also, the two triangles BED and CED, having the same altitude ED, are to each other as their bases BE and CE; hence the two proportions: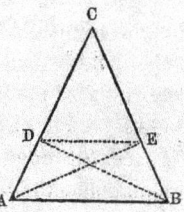

$$ADE : CDE :: AD : CD;$$
$$BED : CDE :: BE : CE.$$

The two triangles ADE and BED are equivalent, having the same base, AB, and the same altitude, since the line DE is parallel with BC; hence, the two proportions above having an antecedent and a consequent of one equivalent to an antecedent and a consequent of the other, the remaining terms are proportional (Prop. V., *Cor.*); hence,

$$AD : CD :: BE : EC;$$
and, by composition, $AD + CD : CD :: BE + EC : EC;$
or, $AC : CD :: BC : CE;$
and, by alternation, $CD : CE :: AD : BE.$

Proposition XXVI. Theorem.

In any triangle, the line which bisects the vertical angle, when produced to the base, divides the base into two parts, which are proportional to the adjacent sides.

Let ABC be any triangle, and let CE bisect the vertical angle C; then will
$$BE : BC :: EA : AC.$$

The angles ACE and BCE are equal by hypothesis; draw AD parallel with CE, and produce it until it intersects the prolongation of BC at D; then will angle D = angle BCE; for they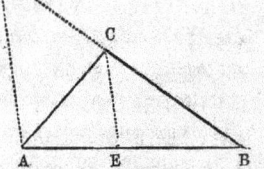
are opposite exterior and interior angles. Also, the angle DAC=ACE, since they are alternate angles; hence, those two angles D and A, in the triangle CAD, equal each other, and the triangle is isosceles. (Prop. XVIII.)

GEOMETRY. 37

In the triangle BAD, since EC is parallel with the base AD, it divides the other two sides proportionally (Prop. XXV.), and we have
$$BE : BC :: EA : CD ;$$
but we have proved the triangle CAD to be isosceles; hence, AC=CD. Substitute, therefore, in the last proportion, AC for its equal CD, and we have, $$BE : BC :: EA : AC.$$

Proposition XXVII. Theorem.

All equiangular triangles are similar, and have their homologous sides proportional.

Let ABC and DEA be two triangles, having the angles, C=E, D=CAB, and B=DAE, then will their homologous sides be proportional, and we shall have

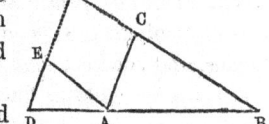

$$BA : AD :: BC : AE ;$$
$$DA : AB :: DE : AC.$$

Place the two triangles so that the side AD shall be the prolongation of the homologous side AB, and produce DE until it intersects the prolongation of BC at F.

Then since the angles EDA and CAB are equal, the lines FD and CA are parallel, for the angles are opposite exterior and interior angles (Prop. XIII. *Cor.*); and since the angles DAE and ABC are equal, the lines BF and AE are parallel, for those angles are opposite exterior and interior angles also; the figure ACEF is therefore a parallelogram, and has its opposite sides equal.

In the triangle BDF, AC being parallel with the base DF, the other two sides are divided proportionally (Prop. XXV.);
and we have
$$BA : AD :: BC : CF. \quad \text{But AE=CF; hence,}$$
$$BA : AD :: BC : AE.$$

In the same manner it may be proved that,
$$DA : AB :: DE : AC.$$

Scholium. It is to be observed, that the homologous or proportional sides are opposite to the equal angles.

Cor. Two triangles are similar, and have their homologous sides proportional, when two angles of the one are respectively equal to two angles of the other; for in that case the third angles must also be equal (Prop. XVI. *Cor.*), and the triangles be equiangular.

Proposition XXVIII. Theorem.

In every convex polygon, the sum of the interior angles is equal to two right angles, multiplied by the number of sides of the given polygon, less two.

Let ABCDEF be any convex polygon, and let diagonals be drawn from any one angle, A, to each of the other angles not adjacent to A; these diagonals will divide the polygon into as many triangles, less two, as the polygon has sides, whatever the number of the sides may be.

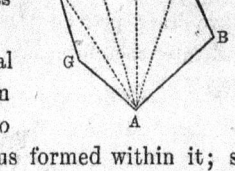

The sum of the angles of every triangle being equal to two right angles (Prop. XVI.), therefore the sum of all the angles of the given polygon will be equal to twice as many right angles as there are triangles thus formed within it; so that, in order to ascertain the entire measure of the angles in any polygon, we have only to multiply two right angles by the number of its sides less two.

Cor. Since $2 \times 2 = 4$, the simplest mode of estimating the measure of the angles of any polygon, is to multiply the entire number of its sides by two right angles, and subtract four from the product.

A *quadrilateral* contains four right angles, since $4 \times 2 = 8$, and $8 - 4 = 4$.

A *pentagon* contains six right angles, since $5 \times 2 = 10$, and $10 - 4 = 6$.

A *hexagon* contains eight right angles, since $6 \times 2 = 12$, and $12 - 4 = 8$.

A *heptagon* contains ten right angles, since $7 \times 2 = 14$, and $14 - 4 = 10$.

Proposition XXIX. Theorem.

When two triangles have the three sides of the one, respectively parallel or perpendicular to the three sides of the other, the two triangles are similar.

In the two triangles ABC and DEF,

First. Let the sides be respectively parallel; namely, let AB be parallel with DE, BC with EF, and AC with DF; then the angles are respectively equal; namely, A=D, B=E, and C=F, since their sides are respectively parallel and lying in the same direction. (Prop. XV.) Hence, their homologous sides are proportional, and they are similar. (Prop. XXVII.)

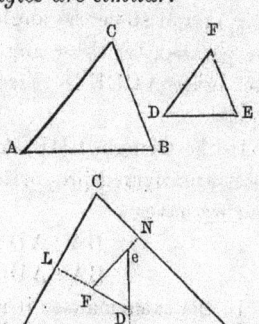

Secondly. Let the sides of the one be respectively perpendicular to the sides of the other; namely, ED perpendicular to AB, FE to BC, and DF to AC; then they will still be equiangular and similar.

In the quadrilateral LADI, the sum of the four interior angles is equal to four right angles (Prop. XXVIII., *Cor.*); but the angles L and I are each right angles, since DL is given perpendicular to AC, and ED to AB; therefore, the sum of the two angles A and LDI is equal to two right angles; but the sum of the angles LDI and LDE equals two right angles (Prop. X.); take away the common angle LDI from each sum, and there remains, A = LDE.

For similar reasons, B=DEF, and C=EFD; hence, the two triangles, being equiangular, have their homologous sides proportional, and are similar. (Prop. XXVII.)

GEOMETRY.

Scholium. The homologous sides are those which are perpendicular or parallel with each other, since they are also those which lie opposite the equal angles.

Proposition XXX. Problem.

To inscribe a square within a given circle.

Let ABCD be the circumference of any circle, and let two diameters, AC and BD, be drawn, intersecting each other at right angles; connect the ends of these diameters by the chords AB, BC, CD, and DA, then will these chords be equal and at right angles with each other, and thus form a perfect, inscribed square.

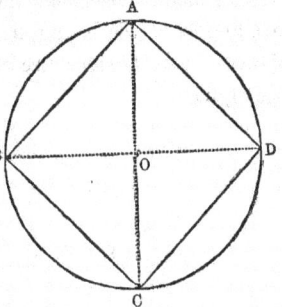

For, AO, BO, DO, and CO are all radii of the same circle, and therefore equal (Def.); the four angles at O are right angles by construction; hence, the four triangles AOB, BOC, COD, and DOA, are equal (Prop. VIII.), and the chords opposite the equal angles at O are also equal. (Prop. IX., *Cor.*)

Again, the angles BAD, ADC, DCB, and CBA are all equal, because they are each composed of two equal angles; and, since their sum equals four right angles (Prop. XXVIII., *Cor.*), each one is a right angle, and the figure ABCD, having four equal sides and four right angles, is a square.

Cor. The arcs embraced within the sides of the equal angles at O, and intercepted by the equal chords, are all equal, since each one is the fourth part of a circumference, or 90°; hence, in the same circle, or in equal circles, equal chords intercept equal arcs, and equal arcs are intercepted by equal chords.

Proposition XXXI. Problem.

To inscribe a regular hexagon and an equilateral triangle within a given circle.

Let ABCDEF be the circumference of any circle. Draw the radii AO and BO, in such a manner that the chord AB, which connects their extremities, shall be equal to the radius itself. This chord will be one side of the regular, inscribed hexagon. For, the triangle ABO, being equilateral, is also equiangular (Prop. XVII., *Cor.*); and the sum of its three angles, being equal to two right angles (Prop. XVI.), each one of its angles is equal to two thirds of a right angle, or 60°, which

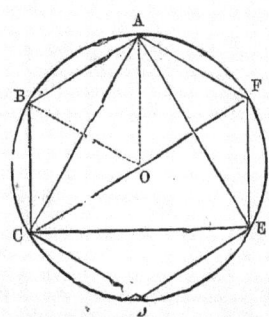

is one sixth of a circumference; hence, the sides of the angle AOB intercept one sixth of the circumference: therefore, the chord AB, applied six times to the circumference, will exactly reach around it, and form a regular hexagon; for the angles of this hexagon will also be equal, since each one is made up of two equal angles, namely, BAO+OAF and ABO+OBC, &c.

After having inscribed the regular hexagon, join the vertices of the alternate angles of the hexagon, and the figure thus formed will be an equilateral triangle; for its sides are chords which intercept equal arcs, and are therefore equal.

PART II.

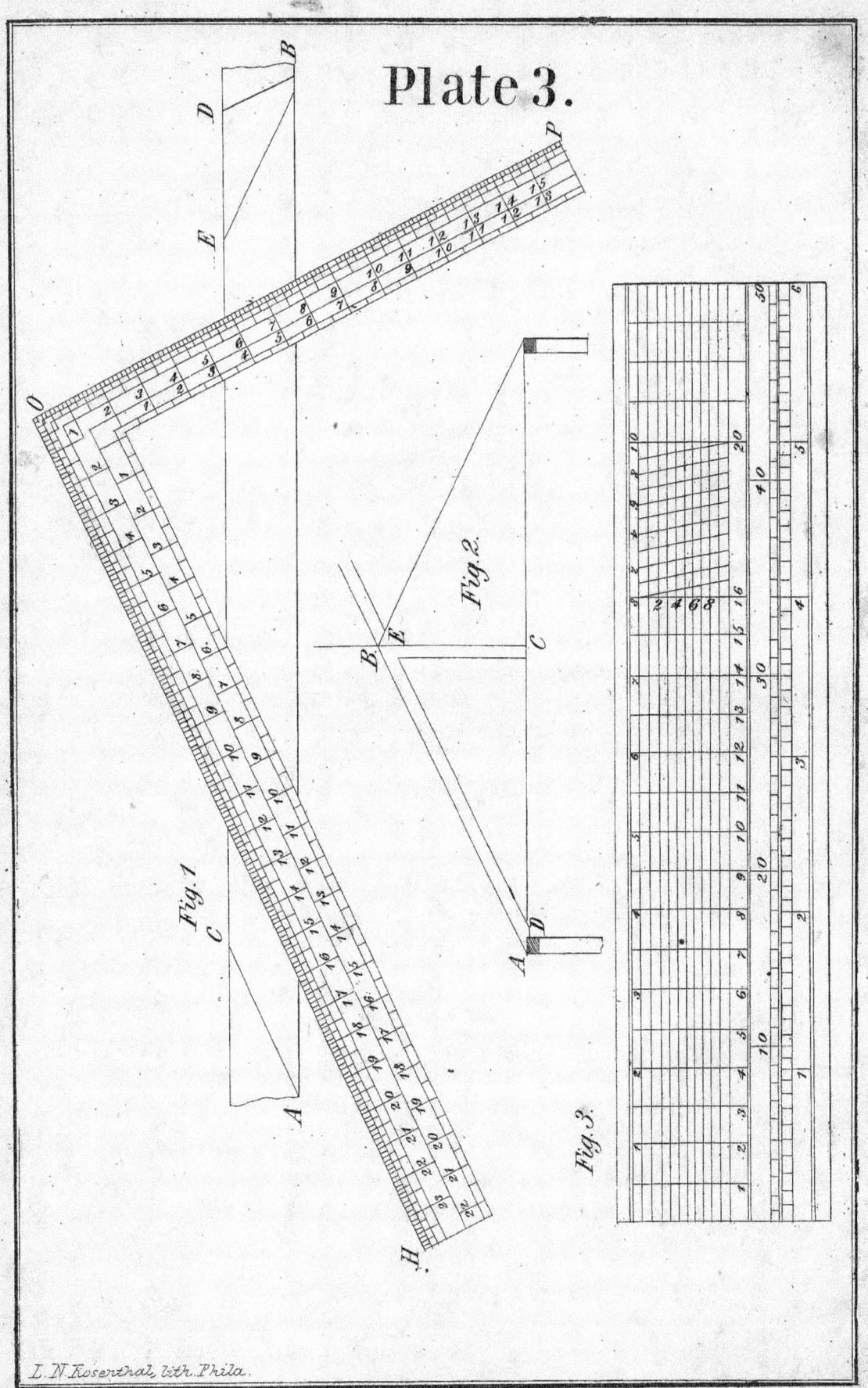

CARPENTRY.

PLATE 3.

THE USE OE THE SQUARE IN OBTAINING BEVELS.

ALTHOUGH the square is one of the first instruments placed in the hands of the practical carpenter, yet there are many experienced mechanics who have never learned all the important uses to which it can be applied. And it is claimed as one of the principal merits of this work, that it teaches the manner of obtaining the bevels of rafters, braces, upper joists, gable-end studding, &c., in the most simple and most accurate manner possible, by the use of the square and scratch-awl alone, without drafts or plans.

The Square Described.

The *common square* is represented in Plate 3, Fig. 1, drawn to the scale of one fourth its size. The point O is called the corner or the heel of the square, the part OH is called the blade, and the part OP the tongue. The blade is 24 inches long; the tongue varies in length in different squares. We commence at the heel to number the inches each way.

Pitch of the Roof.

The bevels of rafters, joists, &c., must, of course, vary with the pitch of the roof. If the roof is designed to have a quarter pitch, which is the most common inclination for a shingle roof, the peak of the roof will be a quarter of the width of the building higher than the top of the plates. Although this is called, among builders, a quarter pitch, yet it would be more simple to call it a half pitch when the roof has two sides, which is most commonly the case, for the true inclination of each its side is 6 inches rise to every foot in width; and in like manner, a third pitch is, in reality, a two thirds inclination to each side of the roof, for it has 8 inches rise to every foot in width.

Bevels of Rafters.

Let AB, in Fig. 1, represent a rafter which is required to be beveled to a quarter pitch. First measure the exact length required, upon the edge AB, (which will be the upper edge of the rafter when it assumes its proper place in the frame,) and let the extreme points A and B be marked. Then place the blade of the square upon the point A at the 12 inch mark, and let the tongue rest upon the edge of the rafter at the 6 inch mark; hold the square firmly in this position, and draw the line AC along the blade: this line will be the lower end bevel. Take the square to the other end of the rafter, and place the 6 inch mark on the tongue upon the point B, still having the blade at the 12 inch mark, and while in this position draw the line DB along the tongue: this line will be the upper end bevel required.

Proceed in a similar manner to mark the bevels for any other pitch, placing the 12 inch mark on the blade upon the point A, and that mark on the tongue which corresponds with the rise of the roof to the foot, on the point B; then the blade of the square will show the lower end bevel, and the tongue the upper end bevel. Thus, if the roof has a pitch of five inches to the foot, let the square be placed at 12 and 5; if the roof has 8 inches rise to the foot, place it at 12 and 8, &c.

The reason of this rule can be explained in few words. In Fig. 2, let C represent the middle point of the line which is drawn from the top of one plate to the top of the other; let AB represent a rafter; and EC the longest gable-end stud, having its longest edge EC directly under the peak of the roof B. The lower end bevel of the rafter rests upon the upper surface of the plate, which is horizontal or level, while its upper end bevel is perpendicular, resting against the upper end of the opposite rafter; so that the upper and lower end bevels of every rafter are always at right angles with each other, whatever the pitch of the roof may be. The tongue of a square is also always at right angles with the blade; and a square can be conceived as having its heel at the point C, its blade resting upon the line AC, and its tongue standing perpendicularly along the line CB. Now let the distance from A to C be supposed to be 1 foot, or 12 inches; then, if the roof is designed to rise 6 inches to the foot, the point B will be 6 inches from C; if it rises 8 inches to the foot, the point B will be 8 inches from

C, &c.; and in all cases the line AC will be one bevel, and the line BC the other.

Bevels of Upper Joists and Gable End Studding.

The bevel of the upper joists is always the same as the lower end bevel of the rafters, and the bevel of the gable end studding is the same as the upper end bevel of the rafters, whatever the pitch of the roof may be. For, in respect to the upper joists, it is to be observed that their lower surfaces rest upon the plates, and their ends are to be beveled to fit the line AB, (Fig. 2), hence the angle BAD, the bevel of the rafter, is identical with that of the end of the joist. The bevel of the end studding is designed to fit it to the lower surface of the rafter; hence the angle DEC is the proper bevel. But DEC=ABC, since they are opposite exterior and interior angles. (Part I. Prop. XII., *Cor.*)

Bevels of Braces.

Proceed in a similar manner to obtain the bevels of braces. When the foot and the head of the brace are to be equally distant from the intersection of the two timbers required to be braced, and when the angle of their intersection is a right angle, then the brace is said to be framed on a *regular run*, and the bevels will be the same at both ends of it, and will always be at an angle of 45°, which is half a right angle, or the eighth part of a circle; and this bevel is obtained from the square by taking 12 on the blade and 12 on the tongue, or any other identical number, the *rise* being equal to the *run*.*

But when the foot and the head of the brace are to be at unequal distances from the intersection of the timbers, the brace is said to be framed on an *irregular run*, and the bevel at one end will be different from that of the other. One rule, however, will answer in all cases. First find the length of the brace from the extreme point of one shoulder to the extreme point of the other, and mark those points as A and B. Then place the blade of the square upon the point A at such a distance from the heel as corresponds with the run of the brace, while the tongue crosses the edge of the brace at that distance from the heel which corresponds with the rise of the brace, and then the blade of the square will show one bevel, and the tongue the other.

For example, a brace is required to be properly beveled for an ir-

---

* For the explanation of those terms, *rise, run*, &c., see the Introduction to the Tables. Part IV.

regular run of 4 by 5 feet. Having found the *length* of the brace (by Table No. 4, or otherwise), and fixed the extreme points of the shoulders, then lay on the square at the 4 and the 5 inch marks, and describe the bevels along the blade and tongue respectively, as in finding the bevels of rafters.

Fig. 3 represents a small ivory rule, drawn full size. It is introduced here for the purpose of showing the manner of taking the measure of hundredths of an inch. It will be perceived that one of the inch spaces of the rule is divided into ten parts, by lines running down diagonally across ten other horizontal lines. Each of the intersections of these lines measures the hundredth part of an inch; the first line measuring tenths, the second twentieths, &c.

Plate 4.

Fig. 1.

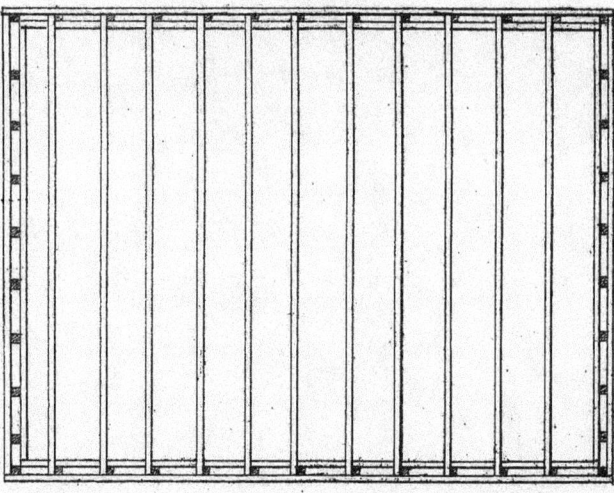

Fig. 2.

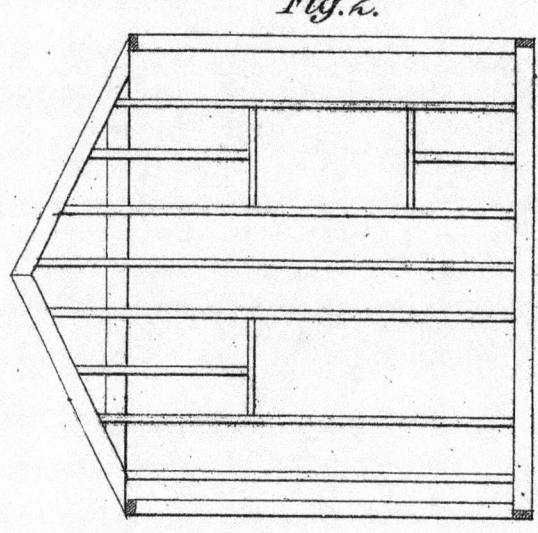

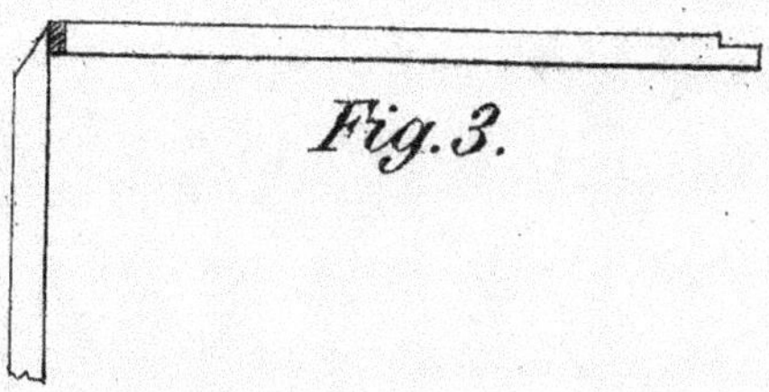

Fig. 3.

L. N. Rosenthal, lith. Phil.ᵃ

PLATE 4.

Balloon Frames.

As Balloon Frames are the simplest of all, they are the first to claim our attention.

The Sills.

Where square timber can conveniently be furnished for sills, it is best to have it; but small buildings can be very well constructed without square sills, even when resting upon blocks only, by using a double set of common joists, with a 2 inch space between them, for the tenons of the studding.

Such a frame, of one story in height, 16 feet long, and 12 feet wide, is represented in Plate 4. For this building, joists which are 2 inches thick and 6 inches wide, will answer. First, for the sills, cut two joists 16 feet long, and two others, 11 feet 8 inches long. Spike them together at the ends in the form of an oblong square, 16 by 12 feet, making the outside rim of the sill.

The Studs.

Next, frame 13 studs for one side of the building; the two corner studs should be 4 inches square, the others 2 by 4. Cut out a 2 inch relish, six inches from the foot of each stud, on the face side, leaving a tenon on the inside of 6 inches long and 2 inches square, as represented in Fig. 3. Then cut off the other end of the studs at 10 feet from the shoulder.

The Plates.

A plate of 2 by 4 stuff, 16 feet long, is now to be nailed flat upon the upper ends of the studs, commencing at the front corner, and taking care to fix them 14 inches apart, or 16 inches from centre to centre. The last space will often be more or less than 14 inches; but it is better to have the odd space all at one end, for the convenience of the plasterers in lathing.

Raising and Plumbing the Frame.

This side of the frame is now ready to be raised. After having prepared the other side in the same manner, that can also be raised.

and the tenons spiked firmly to the inside of the sills. The corners should then be plumbed and securely braced.

The side sills should now be completed by cutting two joists, one for each side, each 15 feet 8 inches long, and framing them for the support of the floor joists by cutting notches into their upper edges 2 inches wide and 2½ inches deep; cutting the first notch 16 inches from the front end, and the next one just 14 inches from that, and so on to the last. After these inside sill-pieces are thus prepared, they should be spiked to their places upon the inside of the tenons of the studs.

The Floor Joists.

The floor joists are to be cut off 11 feet 8 inches long, and their lower corners notched off 2½ inches deep; then they should be fixed in their places in the sills, and also spiked to the studs. By this arrangement the joists are left one inch higher than the sills for the purpose of having the door-sill level with the flooring; the door-sill being 2 inches, and the flooring 1 inch thick

Upper Joists.

The next thing is to frame the upper joists, the rafters and the gable end studs; beveling the ends of each, so as to correspond with the pitch of the roof. The bevel is easily found by the use of the square, as is explained in Plate 3. The upper joists are equal in length to the width of the building. They should be nailed firmly upon the top of the plates, the first one being placed 4 inches from the end of the plate, to leave room for the end studding. The second one should be 14 inches from the first, and the others at the same distance from each other, or 16 inches from centre to centre.

The Rafters.

The exact length of the rafters is found by the use of Table No. 1., Part IV. Look at the left-hand column for the width of the building, and at the top for the rise of the rafter; where those two columns meet in the table, the length of the rafter is found in feet, inches, and hundredths of an inch. In this case, the width of the building is 12 feet, and the rise of the rafter is 6 inches to the foot, or a quarter pitch; therefore the length of the rafter, as given in the table, is (6 : 8.49) 6 feet 8 inches and $\frac{49}{100}$ of an inch. The rule for obtaining these lengths is of perfect accuracy, and is explained in the introduction to the Tables in such a manner that every carpenter can calculate these lengths for himself, from the primary elements, if he chooses. The size of these rafters is 2 by 4.

Gable-End Studs.

The length of the gable-end studding may be found by first calculating the length of the longest one, which stands under the very peak, and then obtaining the lengths of the others from this; or, by first calculating the length of the shortest one next to the corner of the building, and then obtaining the lengths of the others from this.

The length of the middle stud is found by adding to the length of the side studding, the rise of the roof and the thickness of the plate; and deducting from that sum the thickness of the rafter, measured on the upper end bevel. For example, in this building, the length of the side studding from shoulder to shoulder is 10 feet, the rise of the roof is 3 feet, and the thickness of the plate is 2 inches. These all added are 13 feet 2 inches, from which deduct 4.47 inches, or 4½ inches, the thickness of the rafter measured on the upper end bevel, and the result is 12 feet 10½ inches, the length of the middle stud. The next stud, if placed 16 inches from this one, from centre to centre, is 8 inches shorter, since the rise of the roof is 6 inches to 12, or 8 inches to 16. The next one is 8 inches shorter still, and the others in proportion.

If it should be thought preferable to commence by first calculating the length of the shortest one, it can be done. For example, in this building the distance of the inside of the first stud from the outside of the building is 20 inches, the rise of the rafter in running 20 inches back is 10 inches, to which add 2 inches, thickness of the plate, and 10 feet for the length of the side studding, and the sum is 11 feet; from this deduct the thickness of the rafter, at the upper end bevel, 4½ inches, and the result is 10 feet 7½ inches, the length of the shortest stud. The length of the next one is found by adding the rise of the roof in running the distance, that is, if they are 16 inches apart from centre to centre, the difference between them is 8 inches; and so in any other pitch, in the proportion of the *rise* to the *run*.

The end studding having been properly beveled and cut off to the exact lengths required, they can be raised singly and spiked to the sills at the bottom and nailed at the top to the end rafters, and also to the upper joists where they intersect them. After the end studs are all fixed in their positions, the end sills can finally be completed by spiking a joist 11 feet in length to the inside of the studding at each end of the frame.

PLATE 5.

Plate 5 is designed to represent a balloon frame of a building a story and a half high, 16 by 26 feet on the ground, with 12 feet studding. Two end elevations are given, in order to exhibit different styles of roofs. Fig. 2 being a plain roof, of a quarter pitch; and Fig. 3 a Gothic roof, the rafters rising 14 inches to the foot.

Framing the Sills.

Solid timber, 8 inches square, being furnished for the sills of this building, the first business is to frame these. The carpenter will seldom have timber furnished to his hand which is perfectly square throughout its length; by carelessness in hewing, or by the process of seasoning after being hewed, it will most commonly have become irregular and winding.

Work Sides.

Having first selected the two best adjoining sides, one for the upper side and the other for the front, called *work sides*, they should *be taken out of wind* in the following manner.

To take Timber out of Wind.

Plane off a spot on one of the work sides, a few inches from one end, and draw a pencil line square across it; then place the blade of a square upon this line, allowing the tongue to hang down as a plummet, to keep the blade on its edge. Leave the square in this position, and go to the other end of the sill, and place another square upon it, in the same manner; then sight across the two squares, and see if they are level or parallel with each other. If not, make them so, by cutting off the spots under the squares till they become so; then make the other work side square with this one, at these two spots, and draw a pencil mark square across both sides: these marks are called *plumb spots*.

On the upper side of the timber, strike a chalk-line, from one end to the other, at two inches from the front edge; this will be the front line for mortices for studs. On this line measure the length of the

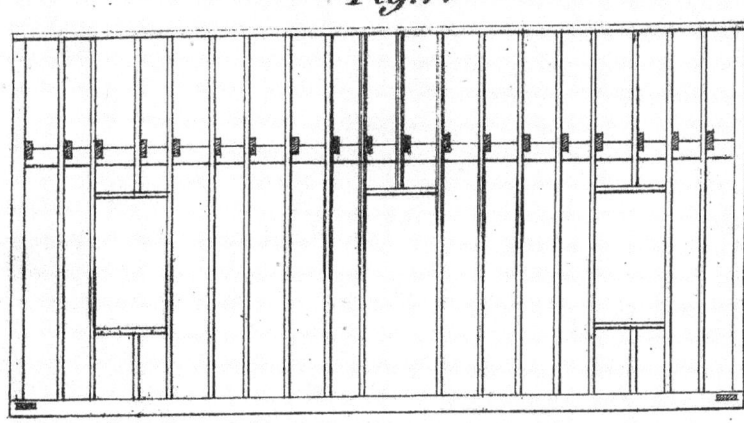

BALLOON FRAMING. 51

sills, and square the ends by it. If the stick is *very irregular*, it should be *counter-hewed*, and the two work sides made square and straight.

Spacing for Windows and Doors.

Next, lay out spaces for windows and doors, leaving a space for the doorway 2½ or 3 inches more than the width of the door; and leave spaces, 7 inches more than the width of the glass, for the windows.

Mortices for the Studs.

Then lay out the mortices for the studding, spacing them as described in Plate 4. The studding on each side of the doors and windows should be 4 inches square, as well as those at the corners of the building. The rest of the studding may be 2×4. The mortices need to be a little more than 2 inches deep, and the tenons 2 inches long.

The lower joists for this building should be 2×8, and 10 inches shorter than the width of the building. They should be placed 16 inches apart, from centre to centre, as already described.

The Gains,

As they are called, for receiving the ends of the joists, should be cut out of the side sills, 4 inches deep and 2 inches square, and 5 inches from the front or outside of the sill. Having framed the sills for the studding and joists, they should next be framed for each other. Make mortices in the ends of the side sills, 2½ inches from the upper surface and 2 inches from the end, 2 inches wide and 5½ inches long. The inside of the sills should be faced off, along the mortice, to within 7½ inches of the work side, in front. The length of the side sills should be the same, of course, as that of the building; but the end sills should be measured from shoulder to shoulder, 15 inches less than the width of the building. Make the tenons of these to correspond with the mortices of those which have just been described.

The Draw Bores.

The draw bores should be 1 inch in diameter, and 1½ inches from the face of the mortice. The draw bore through the tenon should be $\frac{3}{16}$ of an inch nearer the shoulder than that through the mortice, in order to draw the work snugly together.

A Draw Pin.

The proper way to make a *draw pin* for an inch bore is, first, to make it an inch square; then cut off the corners, making it eight-square, then taper it to a point, the taper extending one third the length of the pin. The pin should be about 2 inches longer than the thickness of the timber.

The sills having thus been framed, they can be brought to their places and pinned together, and then the lower joists laid down.

To Support the Upper Joists.

This building being a story and a half high, the upper joists are laid upon a piece of inch board, from 4 to 6 inches wide, which is let into the studs, as seen in the Plate. The bevels and lengths of the rafters are found as already described.

In Fig. 3 the rafters are represented as projecting beyond the plate; this projection may be 3 feet, or more, according to each one's fancy; but whatever it may be, it must be added to the length of the rafter as given in the table, where it is calculated from the upper and outer corner of the plate. The *bevel* will be the same, whatever the additional length may be, as if the rafter did not project at all. In this case, the rafter should be cut out to about one half its width, where it intersects the plate, and must be spiked securely to the plate. The two bevels, at the intersection, will be the same as the upper and lower end bevels, and will make a right angle with each other where they meet at this place.

The *collar beams* can be spiked to the rafters, or they can be dovetailed into them. Both methods are represented in the plate.

Plate 6.
Fig.1.

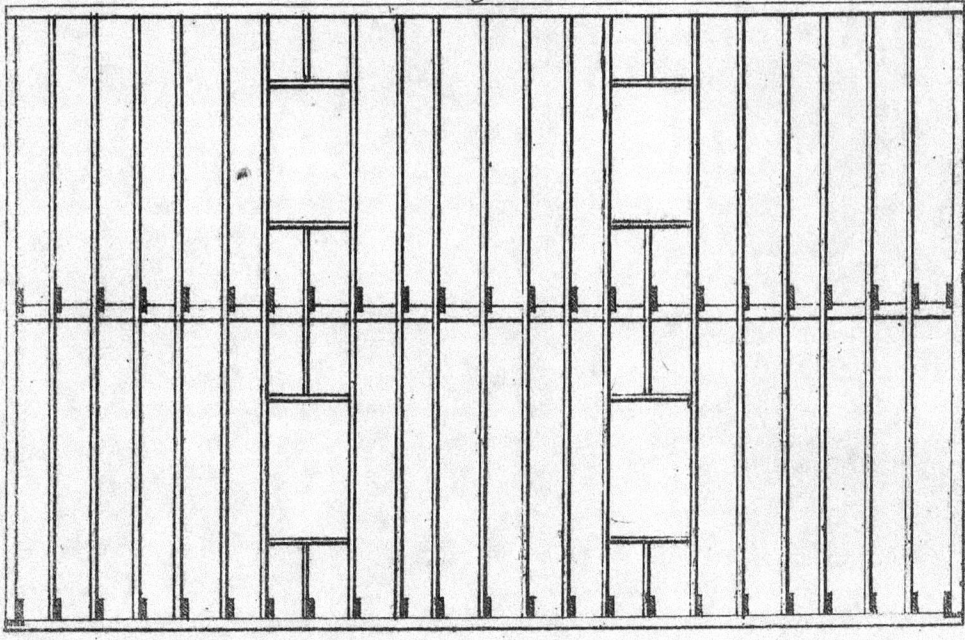

Fig.2.

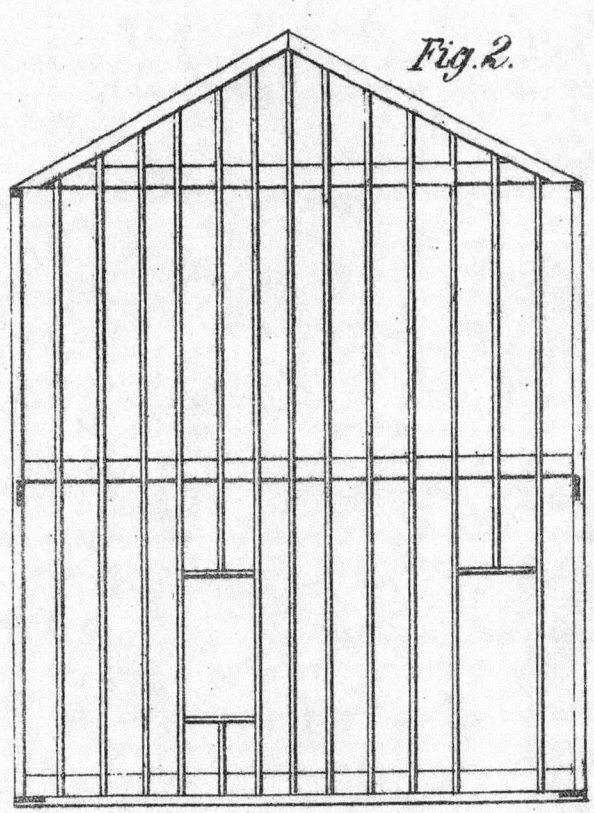

PLATE 6.

Plate No. 6 represents a balloon frame of a two-story building, 18 by 30 feet, with 18 feet studding, to be erected upon a good stone or brick wall. Heavy joists, 3 by 10, are used for sills, with the ends halved together, and fastened with spikes, as represented in the Plate. The lower joists should be 2 by 9 inches, of full length, equal to the width of the building. The lower corners are notched off 3 inches, and they are spiked to the studding. The mortices for the studs should be 1½ inches deep, the studding being 2 by 4. The middle joists are 2 by 9, and arranged as in Plate 5; and the upper, 2 by 7, and arranged as in Plate 4.

Crowning of Joists.

It will almost always happen that one edge of a joist will have become somewhat rounded out, and the other edge rounded in, by the process of seasoning; and it is of much importance, especially in long joists of 18 feet or more, to be careful, in placing the joists in a building, to place the rounding or crowning side up.

Bridging of Joists.

Joists 12 feet long, or over, should also be *bridged* in one or more places, by nailing short pieces of board, 2 or 3 inches wide, in the form of a brace, from the lower edge of one joist to the upper edge of the next one; and then another piece, from the lower edge of this one to the upper edge of the first one; and so on, throughout the whole length of the building: having two braces crossing each other between each joist, beveling the ends so as exactly to fit, which would add very much to the strength of the floor.

Lining or Sheeting Balloon Frames.

After an experience of fifteen years in constructing and repairing balloon-framed buildings, I have found it best to line the frame on the *inside* for three reasons:

FIRST—*the work is more durable.* For, when a frame is lined on the outside, (the common way,) it is very difficult to weather-board it sufficiently tight, to prevent the rain beating in between the siding and the lining, and thus rotting both, since there is so little opportunity there for the moisture to dry out.

SECOND—*the lining is stiffer and warmer.* For, in that case, the lath being but half an inch from the lining-boards,* the mortar is pressed in between every board, making it almost air-tight.

THIRD—*the wall itself is made more solid.* For the mortar being pressed against the lining-boards, is forced both ways, both up and down, forming more perfect clinchers.

---

* When a building is thus lined on the inside, it is best to lath it in the following manner. Single strips of lath are first nailed perpendicularly, sixteen inches apart, upon the lining-boards, and to these the laths for the wall are nailed as usual.

Plate 7.

Fig. 1.

Fig. 2.

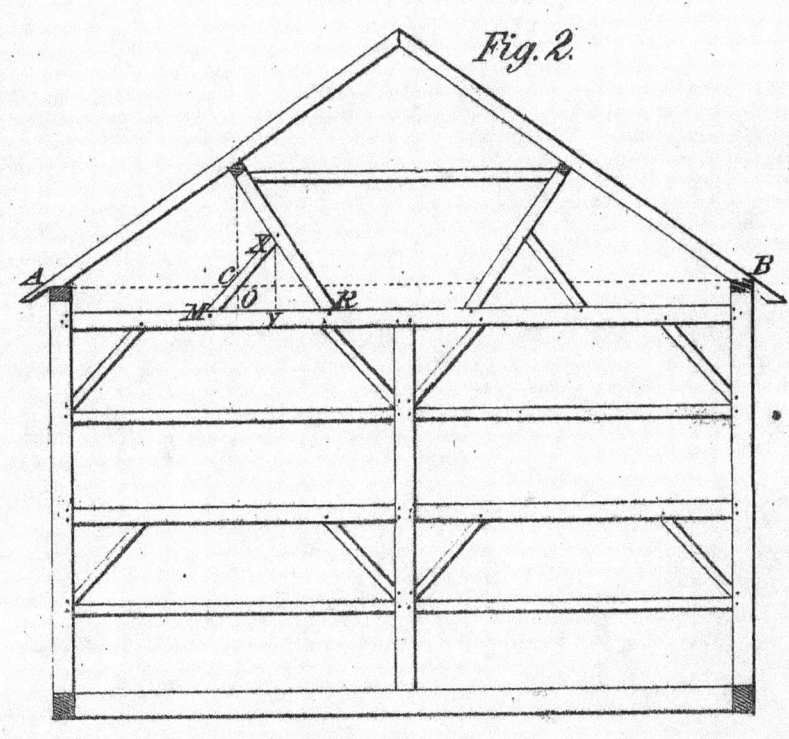

L. N. Rosenthal, lith. Phila.

PLATE 7.

BARN FRAMES.

Plate 7 represents the frame of a barn 30 by 40 feet, and 16 feet high between shoulders.

The sills are 12 inches square;
Posts and large girders, 10 inches square;
Plates and girders over main doors, 8 by 10;
Purlin plates, 6 by 6;
Purlin posts and small girders, 6 by 8;
Braces, 4 by 4; and rafters, 2 by 6.

First proceed to take the timber out of wind, as directed under Plate 5. Frame the sills together as represented in the Plate, the four short sills being framed into the two long ones, having taken care to select the best of the short sills for the ends.

Size of Mortices.

The mortices for the end sills should be 3 by 9 inches, with a relish of $2\frac{1}{2}$ or 3 inches on the outside. The mortices for the middle sills may be 3 by 11 inches. The mortices for the corner posts should be 3 by 7 inches, and for the middle posts, 3 by 9 inches; all the mortices in the sills being 3 inches from the work sides. The general rule for draw bores and draw pins may be stated as follows:— The size of the draw bore should be equal to half the thickness of the tenon, when the tenon is not more than 3 inches thick; but it never need be more than $1\frac{1}{2}$ inches in size, even though the tenon may be more than 3 inches thick. In wide mortices, it is customary to have the tenons secured with two, and sometimes three pins, as represented in the Plate. Let one draw bore be 2 inches from one side of the mortice, and the other 2 inches from the other side, and each one 2 inches from the face of the mortice.

In the tenons, let the draw bores be 2 inches from each side, and about one fourth of an inch, in large tenons, nearer the shoulder than the draw bores of the mortices. Great care should be observed to have the draw bores perfectly plumb; and workmen should be cautioned against making a *push bore*, as it is called, when not plumb.

The posts need not be pinned at the bottom, and the manner of pinning the other tenons is represented in the Plate.

Braces.

The braces are framed on a regular 3 feet run; that is, the brace mortice in the girder is 3 feet from the shoulder of the girder, and the brace mortice in the post is 3 feet below the girder mortice. *Always remember* that the measure for braces and brace mortices is computed to the furthest end, or *toe of the brace*, and the furthest end of the mortice. The mortices for 4 inch braces need to be $5\frac{3}{4}$ inches long, so that the end of the mortice in the post, next the girder, will be 2 feet $6\frac{1}{4}$ inches from the girder, and the end furthest from it will be 3 feet. The bevel of braces on a regular run is always at an angle of 45°, and is the same at both ends of the brace.

Pitch of the Roof.

In this building the roof is designed to have a third pitch; that is, the peak of the roof would be one third the width of the building higher than the top of the plates, provided the rafters were closely fitted to the plates at their outer surfaces, as in Plates Nos. 3, 4, and 6; but it is common in barns, and sometimes in other buildings, as has been already illustrated in Plate 5, Fig. 3, to let the rafters down only half their width upon the plates, allowing them to project beyond the plate, so that in this case the peak of the roof is 10 feet 3 inches above the plates, the pitch being still a third pitch, or 8 inches rise to a foot run. In order to give strength to the mortices for the upper end girders, these girders are framed into the corner post several inches below the shoulders of the post, say 4 inches; the thickness of the plates being 8 inches, it will be perceived that the dotted line, AB, drawn from the outer and upper corner of one plate to the outer and upper corner of the other, is just 1 foot higher than the upper surface of the girder; and that the peak of the roof is 11 feet 3 inches above this girder. The length and bevels of the rafters can be found as already described in Plate 3 and Table 1.

Purlins.

The purlin plates should always be placed under the middle of the rafters; and the purlin posts, being always framed square with the purlin plates, the bevel at the foot of these posts will always be the same as the upper end bevel of the rafters;* also, the bevel at each end of the gable-end girder will be the same, since—the two girders being parallel, and the purlin post intersecting them—the *alternate angles* are equal. (Prop. XII., *Cor.*, p. 29.) The length of the gable-end girder will be equal to half the width of the building, less 18 inches; 6 inches being allowed for half the thickness of the purlin posts, and 6 inches more at each end for bringing it down below the shoulders of the posts.

Length of the Purlin Posts.

In order to obtain the length of the purlin posts, let the learner pay particular attention to the following explanation of Fig. 2. Let the point P represent the middle point of the rafter, and let the dotted line PO be drawn square with AB; then will AC be the $\frac{1}{4}$ of AB, or $7\frac{1}{2}$ feet, and PC, half the rise of the roof, will be 5 feet, and PO 6 feet. The purlin post being square with the rafter, and PO being square with AB, we can assume that PR would be the rafter of another roof of the same pitch as this one, provided PO were half its width, and OR its rise; and then, since we know the length of PO, the length of PR could also be found by the rafter table (No. 1, Part IV.), as follows:—Width of building, 12 feet; rise of rafter, $\frac{1}{3}$ of 12, or 4 feet; hence, length of rafter, or PR, equals 7 feet $2\frac{52}{100}$ inches; from this deduct half the width of the rafter and the thickness of the purlin plate, or 9 inches, and we have, 6 feet $5\frac{52}{100}$ inches as the length of the purlin post, from the shoulder at the top to the middle of the shoulder at the foot.† This demonstration determines also the place of the purlin post mortice in the girder; for AC being $7\frac{1}{2}$ feet, and OR being 4 feet, by adding these together, we find the point R, the

---

* This fact is capable of a geometrical demonstration; for the triangle POR is similar to the triangle ACP; the side PR in one, being perpendicular to the side AP in the other, the side PO being also perpendicular to AC, and the side RO perpendicular to PC. (Part I., Prop. XXIX.) Hence, the angles opposite the perpendicular sides are equal; and we have angle APC, which is the same as the upper end bevel of the rafter —being parallel with it—equal to PRC, the angle formed by the purlin post and the girder at their intersection at R.

† The following geometrical demonstration of the above proposition is subjoined. In the two similar triangles ACP and POR, the sides about the equal angles are proportional (Def. 31); and we have, CP : AC : : OR : OP; but CP is $\frac{2}{3}$ of AC; consequently, OR is $\frac{2}{3}$ of OP. But OP equals 6 feet; hence, OR equals 4 feet. Again, the triangle POR being right-angled at O, then $PO^2 + OR^2 = PR^2$. $4^2 = 16$, and $6^2 = 36$; $36 + 16 = 52$, and $\sqrt{52}$ ft. $= 7$ ft. 2.52 in., as above.

middle of the mortice, to be 11½ feet from the outside of the building; and the length of the mortice being 7¼ inches, the distance of the end of the mortice, next the centre of the building, is 11 feet 9⅝ inches from the outside of the building.

Purlin Post Brace.

The brace of the purlin post must next be framed, and also the mortices for it, one in the purlin post and the other in the girder. The length of the brace and the lower end bevel of it will be the same as in a regular 3 feet run; and the upper end bevel would also be the same, provided the purlin post were to stand perpendicular to the girder; but, being square with the rafter, it departs further and further from a perpendicular, as the rafter approaches nearer and nearer toward a perpendicular; and the upper end bevel of the brace varies accordingly, approaching nearer and nearer to a right angle as the bevel at the foot of the post, or, what is the same thing, the upper end bevel of the rafter departs further and further from a right angle. Hence, *the bevel at the top of this brace is a* COMPOUND BEVEL, *found by adding the lower end bevel of the brace to the upper end bevel of the rafter.** (See Plate 8.)

Purlin Post Brace Mortices.

In framing the mortices for the purlin post braces, it is to be observed, also, that if the purlin post were perpendicular to the girder, the mortices would each of them be 3 feet from the heel of the post; but as the post always stands back, so the distance will always be more than 3 feet from the heel of the post; and the sharper the pitch of the roof, the greater this distance will be. Hence the true distance on the girder for the purlin post brace mortice is found by adding to 3 feet the rise of the roof in running 3 feet; which, in this pitch of 8 inches to the foot, is 2 feet more, making 5 feet, the true distance of the furthest end of the mortice from the heel of the purlin post.

The place in the purlin post for the mortice for the upper end of the brace may be found from the rafter table, by assuming that Rx

---

* This proposition is capable of demonstration, thus: The angle PzM equals the sum of the angles MRx and xMR, since PzM is the exterior angle of the triangle MRx, formed by producing the base Rx in the direction RxP. (See Prop. XVI., *Cor.*) But the angle PzM is the upper end bevel of the purlin post brace; therefore, it is equal to the sum of the two bevels, one at the foot of the brace and the other at the foot of the post, as above.

BARN FRAMING.

would be the rafter of another roof of the same pitch as this one, if xy were half the width, and yR the rise. For then, since xy equals 3 feet, we should have width of building equal 6 feet, rise of rafter, one third pitch, gives yR equal 2 feet; and hence xR would equal 3 feet 7.26 inches, the true distance of the upper end of the mortice from the heel of the purlin post.*

---

* The same proposition is demonstrated by Geometry, as follows: xy being parallel with PO, the two triangles RPO and Rxy are similar, (Geom., Prop. XXIX), hence the sides opposite the equal angles are proportional, and we have Rx : RP : : xy : PO. But we have already found PO to equal 6 feet, and xy equal to 3 feet, and RP equal to 7 feet 2.52 inches. Hence,
 6 : 3 : : 7 ft. 2.52 in. : 3 ft. 7.26 in. Answer as above.

PLATE 8.

UPPER END BEVEL OF PURLIN POST BRACES.

Plate 8 is designed to illustrate the manner of finding the upper end bevel of purlin post braces, to which reference is made from the preceding Plate.

In Fig. 1, let AB represent the extreme length of the brace from toe to toe, the bevel at the foot having been already cut at the proper angle of 45 degrees. Draw BC at the top of the brace, at the same bevel; then set a bevel square to the bevel of the upper end of the rafter, and add that bevel to BC, by placing the handle of the square upon BC and drawing BD on the tongue. This is the bevel required.

Fig. 2 shows another method of obtaining the same bevel. Let the line AB represent the bevel at the foot of the brace, drawn at an angle of 45 degrees. Draw BD at right angles with AB, and draw BC perpendicular to AD, making two right-angled triangles. Then divide the base of the inner one of these triangles into 12 equal parts, for the rise of the roof. Then place the bevel square upon the bevel AB, at B, and set it to the figure on the line CD, which corresponds with the pitch of the roof. This will set the square to the bevel required for the top of the brace. In this figure the bevel is not marked upon the brace, but the square is properly set for a pitch of 8 inches to the foot, or a one third pitch. The square can now be placed upon the top of the brace, and the bevel marked.

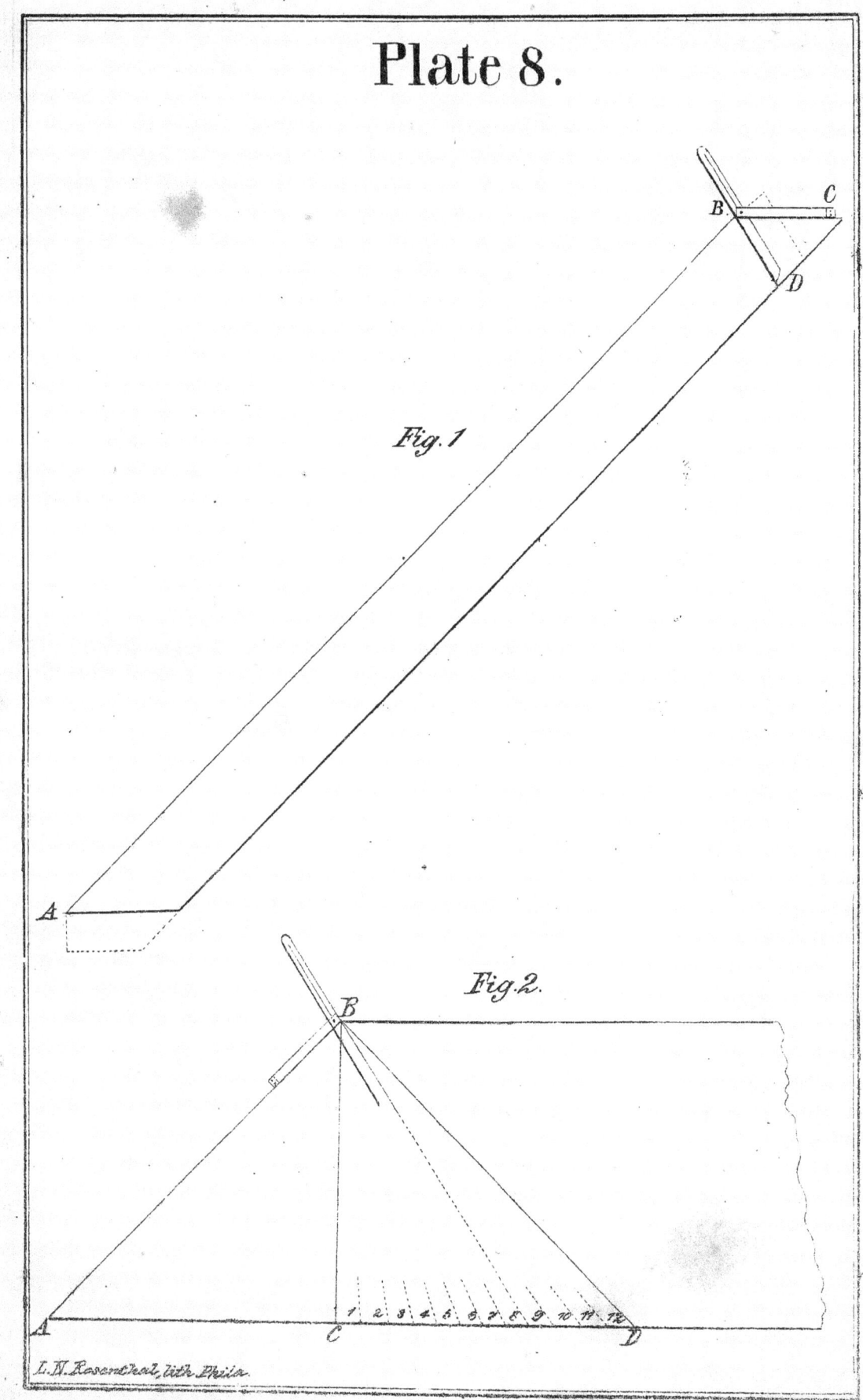

Plate 9.

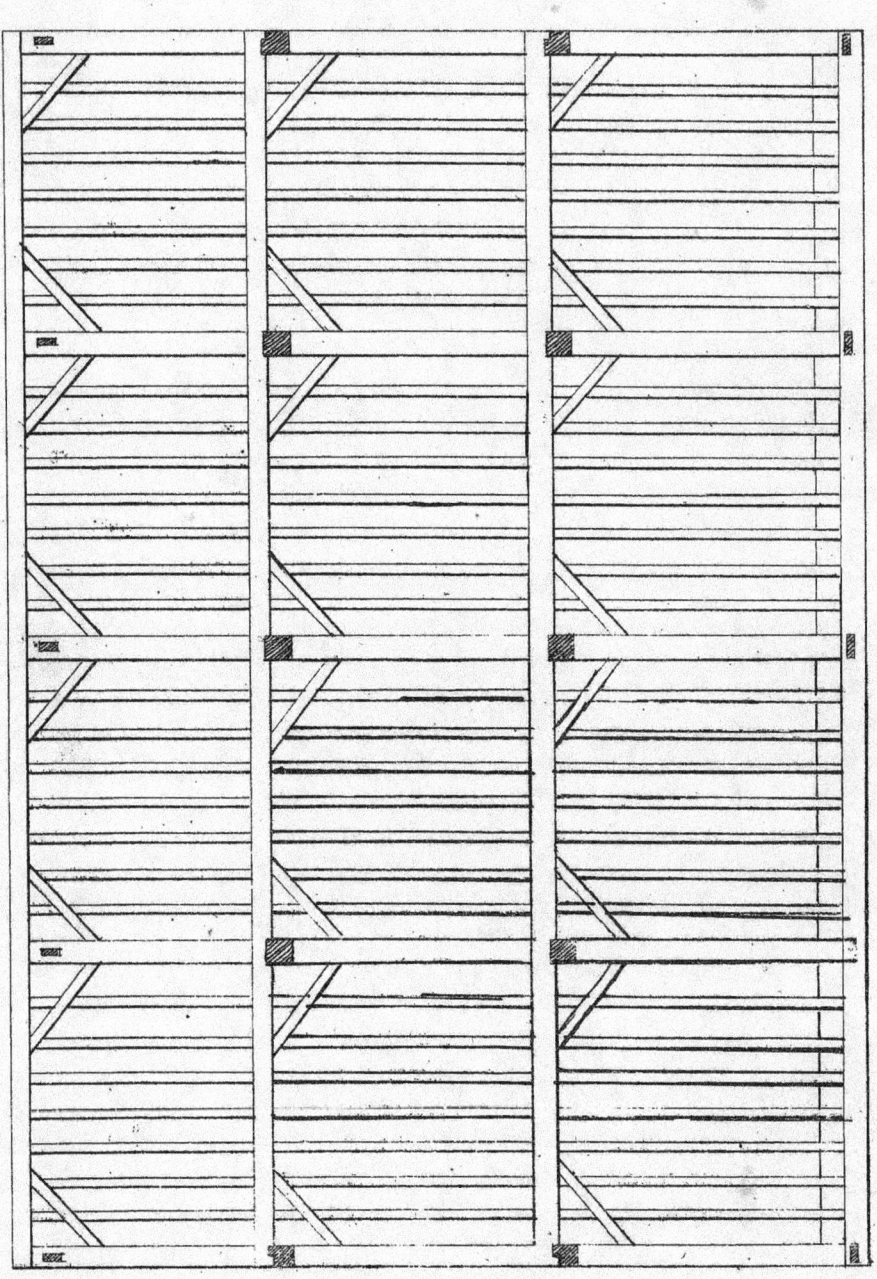